Organic Synthesis via Transition Metal-Catalysis

Organic Synthesis via Transition Metal-Catalysis

Editor

Bartolo Gabriele

MDPI • Basel • Beijing • Wuhan • Barcelona • Belgrade • Manchester • Tokyo • Cluj • Tianjin

Editor
Bartolo Gabriele
Department of Chemistry and
Chemical Technologies
University of Calabria
Rende, CS
Italy

Editorial Office
MDPI
St. Alban-Anlage 66
4052 Basel, Switzerland

This is a reprint of articles from the Special Issue published online in the open access journal *International Journal of Molecular Sciences* (ISSN 1422-0067) (available at: https://www.mdpi.com/journal/molecules/special_issues/organic_synthesis_transition_metal_catalysis).

For citation purposes, cite each article independently as indicated on the article page online and as indicated below:

Lastname, A.A.; Lastname, B.B. Article Title. *Journal Name* **Year**, *Volume Number*, Page Range.

ISBN 978-3-0365-3233-2 (Hbk)
ISBN 978-3-0365-3232-5 (PDF)

© 2022 by the authors. Articles in this book are Open Access and distributed under the Creative Commons Attribution (CC BY) license. The book as a whole is distributed by MDPI under the terms and conditions of the Creative Commons Attribution-NonCommercial-NoDerivs (CC BY-NC-ND) license.

Contents

About the Editor vii

Garazi Urgoitia, Maria Teresa Herrero, Fátima Churruca, Nerea Conde and Raul SanMartin
Direct Arylation in the Presence of Palladium Pincer Complexes
Reprinted from: *Molecules* **2021**, *26*, 4385, doi:10.3390/molecules26144385 1

Mieko Arisawa and Masahiko Yamaguchi
Rhodium-Catalyzed Synthesis of Organosulfur Compounds Involving S-S Bond Cleavage of Disulfides and Sulfur
Reprinted from: *Molecules* **2020**, *25*, 3595, doi:10.3390/molecules25163595 21

Jérémy Ternel, Adrien Lopes, Mathieu Sauthier, Clothilde Buffe, Vincent Wiatz and Hervé Bricout et al.
Reductive Hydroformylation of Isosorbide Diallyl Ether
Reprinted from: *Molecules* **2021**, *26*, 7322, doi:10.3390/molecules26237322 57

Joanna Palion-Gazda, André Luz, Luis R. Raposo, Katarzyna Choroba, Jacek E. Nycz and Alina Bieńko et al.
Vanadium(IV) Complexes with Methyl-Substituted 8-Hydroxyquinolines: Catalytic Potential in the Oxidation of Hydrocarbons and Alcohols with Peroxides and Biological Activity
Reprinted from: *Molecules* **2021**, *26*, 6364, doi:10.3390/molecules26216364 69

Lucia Veltri, Roberta Amuso, Marzia Petrilli, Corrado Cuocci, Maria A. Chiacchio and Paola Vitale et al.
A Zinc-Mediated Deprotective Annulation Approach to New Polycyclic Heterocycles
Reprinted from: *Molecules* **2021**, *26*, 2318, doi:10.3390/molecules26082318 93

Zhe Chang, Tong Ma, Yu Zhang, Zheng Dong, Heng Zhao and Depeng Zhao
Synthesis of Functionalized Indoles via Palladium-Catalyzed Cyclization of *N*-(2-allylphenyl) Benzamide: A Method for Synthesis of Indomethacin Precursor
Reprinted from: *Molecules* **2020**, *25*, 1233, doi:10.3390/molecules25051233 107

Bao Wang, Xu Han, Jian Li, Chunpu Li and Hong Liu
Sulfoximines-Assisted Rh(III)-Catalyzed C–H Activation and Intramolecular Annulation for the Synthesis of Fused Isochromeno-1,2-Benzothiazines Scaffolds under Room Temperature
Reprinted from: *Molecules* **2020**, *25*, 2515, doi:10.3390/molecules25112515 117

About the Editor

Bartolo Gabriele

Prof. Bartolo Gabriele completed his PhD in "Chemical Sciences" in 1994 at the University of Calabria (UNICAL). In 1997, he joined Professor Ronald Breslow's group at Columbia University (New York) for 1 year as a NATO-CNR fellow. In 1998, he returned to UNICAL, where he was promoted to a Full Professor in 2006. His current research interests include the synthesis of high value-added molecules by catalytic techniques and the development of new materials for advanced applications. He has published more than 215 articles in international peer-reviewed journals and more than 10 chapters in international books, and has 20 industrial patents. H-index: 49; total citations: 7394 (Scopus). He received the "Research Award" in 2013 from the Italian Chemical Society, in the field "Synthetic Organic Chemistry". He has been involved, as a coordinator or participant, in many national and international projects.

Review

Direct Arylation in the Presence of Palladium Pincer Complexes

Garazi Urgoitia, Maria Teresa Herrero, Fátima Churruca, Nerea Conde and Raul SanMartin *

Department of Organic and Inorganic Chemistry, Faculty of Science and Technology, University of the Basque Country (UPV/EHU), 48940 Leioa, Spain; garazi.urgoitia@ehu.eus (G.U.); mariateresa.herrero@ehu.eus (M.T.H.); fatima.churruca@ehu.eus (F.C.); nerea.conde@ehu.eus (N.C.)
* Correspondence: raul.sanmartin@ehu.eus; Tel.: +34-94-6015435

Abstract: Direct arylation is an atom-economical alternative to more established procedures such as Stille, Suzuki or Negishi arylation reactions. In comparison with other palladium sources and ligands, the use of palladium pincer complexes as catalysts or pre-catalysts for direct arylation has resulted in improved efficiency, higher reaction yields, and advantageous reaction conditions. In addition to a revision of the literature concerning intra- and intermolecular direct arylation reactions performed in the presence of palladium pincer complexes, the role of these remarkably active catalysts will also be discussed.

Keywords: direct arylation; palladium; pincer complexes

Citation: Urgoitia, G.; Herrero, M.T.; Churruca, F.; Conde, N.; SanMartin, R. Direct Arylation in the Presence of Palladium Pincer Complexes. *Molecules* **2021**, *26*, 4385. https://doi.org/10.3390/molecules26144385

Academic Editor: Bartolo Gabriele

Received: 23 June 2021
Accepted: 16 July 2021
Published: 20 July 2021

Publisher's Note: MDPI stays neutral with regard to jurisdictional claims in published maps and institutional affiliations.

Copyright: © 2021 by the authors. Licensee MDPI, Basel, Switzerland. This article is an open access article distributed under the terms and conditions of the Creative Commons Attribution (CC BY) license (https://creativecommons.org/licenses/by/4.0/).

1. Introduction

Many biologically active compounds contain (hetero)biaryl frameworks. In fact, this pharmacophore core is found in a number of currently prescribed or clinically tested drugs, including several employed in the therapy against cancer or infertility, or antifungal, anti-inflammatory, anti-hypertensive and antibiotic drugs, inter alia (Figure 1) [1–6]. In addition to some agrochemical compounds, relevant materials such as liquid crystal displays and molecular switches comprise the (hetero)biaryl motif [7–10]. Among the methods developed for the preparation of (hetero)biaryl compounds, Ullmann, Scholl, and Gomberg–Bachmann reactions are considered to be classical strategies [11–13], whereas palladium or nickel-catalyzed cross-coupling reactions (Suzuki–Miyaura [14], Kumada [15], Stille [16] and Negishi couplings [17]) were discovered at the end of the 20th century and have been extensively utilized due to the large substrate scope and milder conditions involved. Nevertheless, pre-activated or functionalized coupling partners are required for the latter cross-coupling reactions, as (hetero)aryl halides or pseudohalides are coupled with organometallic reagents (organoboron, organomagnesium, organotin, organozinc compounds, respectively). Additional synthetic steps are therefore needed, and the coupling reaction itself often involves the generation of stoichiometric amounts of metal waste. In order to avoid these inconveniences, new methods for (hetero)aryl-aryl bond formation have been devised [18–24]. In this regard, direct arylation reactions have arisen as a promising alternative to the above cross-coupling strategies. Different names have been applied to define the coupling of a simple (hetero)arene with an aryl halide or pseudohalide. Among them, C–H (bond) activation, cross-dehalogenative coupling, C–H (bond) functionalization, and catalytic direct arylation are generally employed. Although pioneering reports on the use of alternative metals for direct arylation have been published (Cu [25], Fe [26], Ni [27], Ir [28] or Co [29]), second-row transition metals in low oxidation states (Rh [30–33], Ru [34–38], and especially Pd [39–43]) are preferred as catalysts for these cross-dehalogenative couplings.

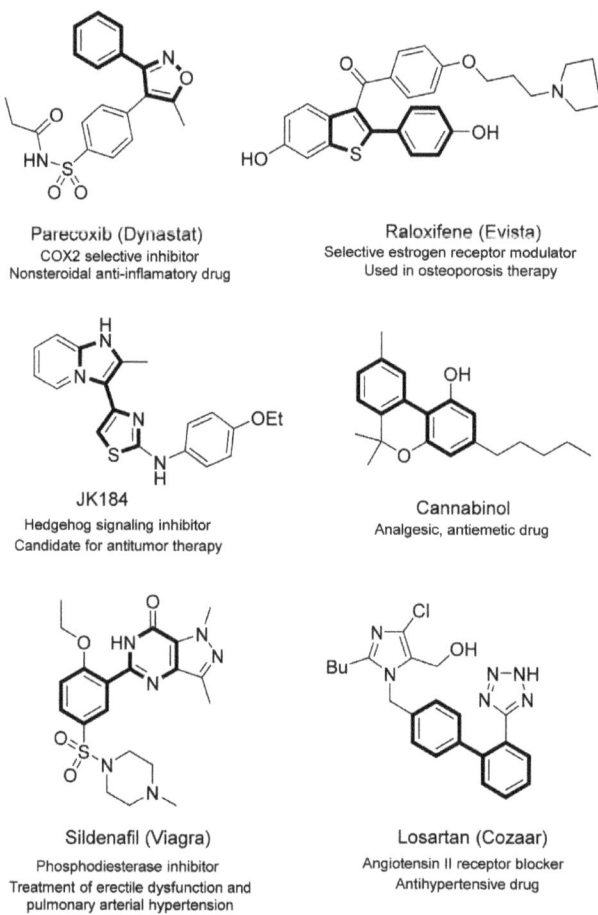

Figure 1. Examples of important bi(hetero)aryl-containing compounds.

The ligands required usually depend on the nature of the haloarene coupling partner. Thus, monodentate triaryl phosphines (e.g., PPh$_3$ and P(o-Tol)$_3$) are typically used for more reactive iodoarenes. Arylation with bromo(hetero)arenes can be also carried out with the same phosphines, although with some substrates better results have been obtained using sterically crowded, electron-rich trialkylphosphanes and biphenylphosphanes [44–51]. The use of chloroarenes in most cross-coupling reactions is often hampered by the more difficult oxidative addition step [52,53]. Therefore, the palladium-catalyzed direct arylation of chloroarenes is usually carried out in the presence of the above trialkyl- and biphenylphosphanes or N-heterocyclic carbenes (NHC) as ligands. Jeffery's ligand-free conditions have also been successfully used in this field [54–61]. Catalyst loading generally ranges from 1 to 20 mol%.

Alkali carbonates (K$_2$CO$_3$, Cs$_2$CO$_3$), carboxylates (KOAc, CsOPiv) and ^{t}BuOK are the bases which are usually employed, although in some cases, bases such as DBU and Et$_3$N have been described. In addition to regenerate the active catalyst, it has been proposed that those bases take part in the formation of diarylpalladium(II) species [62–64]. In part due to the higher solubility in organic solvents, Cs$_2$CO$_3$ and CsOPiv have provided better results in some cases. As for solvents, although non-polar toluene and xylene have been employed, N,N-dimethylformamide (DMF), N,N-dimethylacetamide (DMA), acetonitrile,

N-methylpyrrolidone (NMP) and dimethylsulfoxide (DMSO) are the commonly used polar aprotic solvents. Heating at temperatures ranging from 100 °C to 140 °C for several hours or days is generally required [65]. Interestingly, a recent report by Albéniz and co-workers demonstrated the beneficial and non-innocent role of alternative solvents such as pinacolone [66].

Several mechanisms have been proposed to explain the direct arylation process. After an initial oxidative addition step, the postulated mechanisms diverge in different pathways. Thus, an electrophilic aromatic substitution-type process might take place [67–69], or a concerted termolecular electrophilic substitution [70], or base-assisted intramolecular electrophilic-type substitution [71], or a σ-bond metathesis [72,73], a single electron transfer (SET) [74], or a carbometallation process followed by a β-hydride elimination [75,76], a concerted metalation deprotonation (CMD) [77], or a C–H bond oxidative addition [78–80]. Alternatively, in concordance with ruthenium-catalyzed arylations [81,82], a mechanistic pathway based on an initial palladium-catalyzed C–H bond activation has been proposed. As shown in Scheme 1, interaction between the Pd(II) complex and the arene would result in the generation of arylpalladium complex $Ar^1Pd(II)L$, which, upon transmetallation with Ar^2X, forms intermediate $Ar^1Pd(II)Ar^2$. After reductive elimination of the latter complex with the release of Pd(0) species and Ar^1-Ar^2, the catalyst would be regenerated by oxidation to Pd(II) [83].

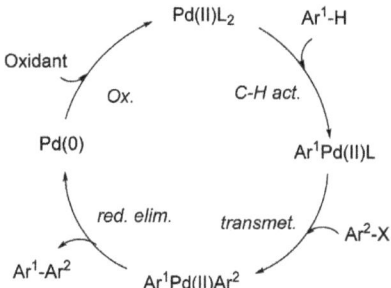

Scheme 1. Alternative pathway to explain palladium-catalyzed direct arylation.

Regioselectivity is often controlled by the electronics of the arene in which C–H functionalization takes place, by the relative C–H acidity, and by the presence of directing groups (nitrogen- or oxygen-coordinating groups, tethering groups, or intramolecular arylations) [84]. As examples of regioselective direct arylation based on the presence of directing groups, Kim and co-workers described the palladium-catalyzed C-8 arylation of dihydroisoquinolones [85]. A regioselective C-3 phenylation of 1-methylquinolin-4(1H)-one was reported by Choi et al. (Scheme 2) [86], and Hartwig's group presented the regioselective arylation of a number of mono- and disubstituted arenes using synergistic silver and palladium catalysis [87].

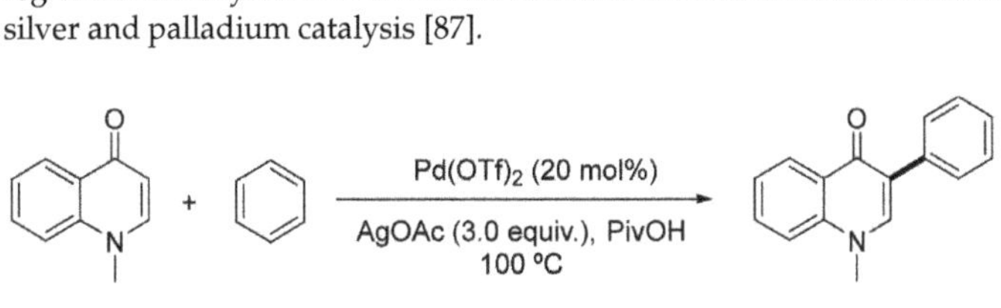

Scheme 2. Regioselective C-3 phenylation of 1-methylquinolin-4(1H)-one.

Direct arylation has been a tool for the construction of several natural products, polycyclic aromatic hydrocarbons (PAHs) and other chemically relevant compounds [88–90]. Indeed, the structure of several biologically relevant lactones prepared by direct arylation, including *Defucogilocarcins M* and *E* [91,92], and intermediates for the syntheses of

Dioncophylline A and *Mastigophorene B* [93–95], are shown in Figure 2. Structurally related *Arnottin I* was prepared by Harayama and co-workers by direct arylation using Pd(OAc)$_2$ (10 mol%) [96]. The preparation of several korupensamines, i.e., a family of naphthyltetrahydroisoquinoline alkaloids with antimalarial activity, was reported by Bringmann. An intramolecular direct arylation provided the lactone-bridged biaryl displayed in Figure 2, which was atroposelectively cleaved. Arylation employing Herrmann's catalyst [97] provided a good yield of the required biaryl intermediate, whereas poor results were obtained by using Pd(OAc)$_2$/PPh$_3$. Accordingly, palladacycles can be useful, more efficient palladium sources for direct arylation [98].

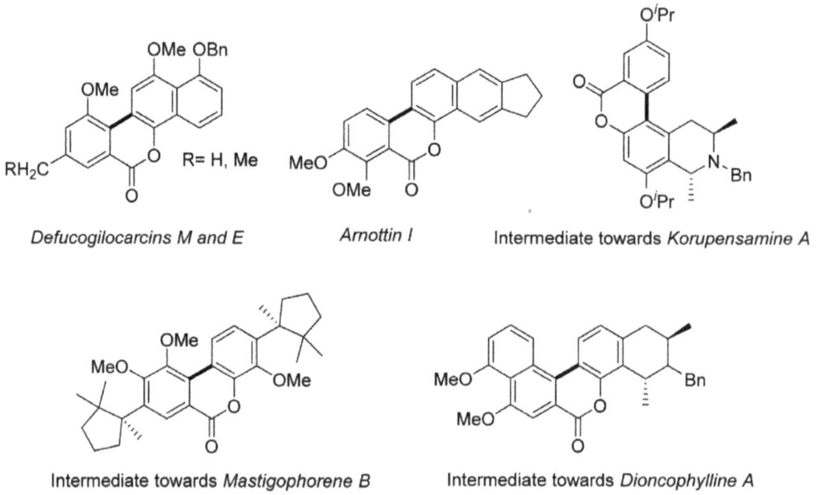

Figure 2. Natural products and synthetic precursors obtained via direct arylation.

Bowl-shaped PAHs are typical synthetic precursors for the access to fullerenes. These fullerene fragments have been prepared in moderate to poor yields by flash vacuum pyrolysis. Moreover, as a result of the harsh reaction conditions needed, only a limited substrate scope is achieved, and this method is difficult to scale up. Following a pioneering report by Rice and co-workers [99], a number of PAHs including bowl-shaped fullerene fragments have been successfully synthesized by the intramolecular direct arylation of *o*-functionalized biaryl and benzophenanthrene derivatives. High yields and a good tolerance of functional groups were achieved (Scheme 3) [100,101].

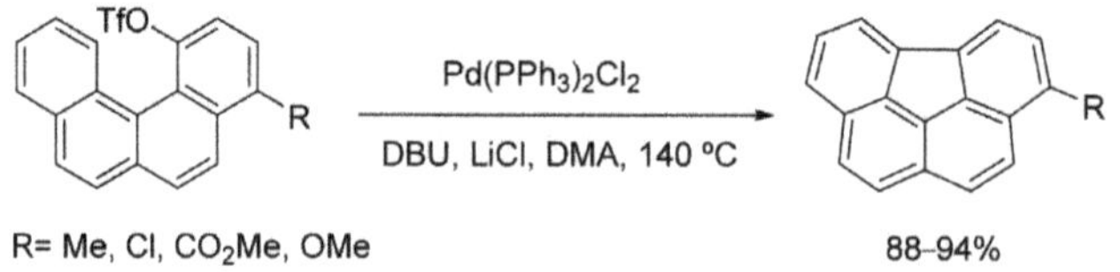

Scheme 3. Intramolecular direct arylation for the synthesis of PAHs.

Moulton and Shaw [102] reported the first examples of pincer complexes in 1976. High thermal, air and moisture stability were exhibited by palladium pincer complexes due to the tight coordination of the tridentate ligand to palladium. Although initially most of these complexes were symmetrical, non-palindromic ligands with hard and soft donor atoms have also been incorporated, thus providing a whole variety of structural designs. Depending on the latter structural features, interaction with substrates and/or the stabilization of reaction intermediates can be facilitated [103,104]. In fact, an increasing number of reports on the application of these terdentate complexes as catalysts or pre-catalysts for a number of synthetic transformations have been published, including recent papers on cross-coupling reactions catalyzed by pincer compounds [105–120]. In this regard, inter- and intramolecular direct (hetero)arylations have been carried out in the presence of several pincer complexes. This review will cover the literature on the use of palladium pincer complexes for the C–H functionalization of (hetero)arenes with (hetero)aryl halides. A brief summary of the reaction scope, and in some cases, proposals on the role of the pincer complex, will be discussed.

2. Palladium(II) Complexes with Phosphine-Containing Pincer Ligands

Two palladium PCP and PCN complexes were tested as catalysts for the direct access to pyrazolo(benzo)thienoquinolines. This approach involved the intramolecular heteroarylation of 1-aryl-5-(benzo)thienylpyrazoles. The authors confirmed the excellent performance of the above complexes in comparison with commercially available $Pd(OAc)_2$. Indeed, good to excellent yields for the target tetra- and pentacyclic compounds were obtained by using a relatively low amount (1 mol%) of phosphinite- and phosphinoamide-based PCP and PCN complexes. As for $Pd(OAc)_2$, a significantly higher 10 mol% was required to catalyze the same reaction under Jeffrey's ligand-free conditions, and even then, the yields obtained were lower in all cases. However, no clear differences were found between the catalytic ability of symmetric **PCP** and non-symmetric **PCN** complexes (Scheme 4). In addition to this palladium-catalyzed intramolecular heteroarylation, the authors also reported the intermolecular regioselective C-5 arylation of simple 1-substituted thiophenes with an equimolecular amount of bromobenzene under similar conditions [121].

Punji and coworkers reported the intermolecular direct C-2 arylation of benzothiazoles with aryl iodides in the presence of a palladium PCN pincer complex (**POCN**). 5-Aryloxazoles were also regioselectively arylated at the C-2 position under the same reaction conditions, which involved the use of cesium carbonate as a base and a slight excess of the iodoarene (1.5 equiv.) in DMF at 120 °C. Catalyst loading was optimized at 0.5 mol%, although it was necessary to add CuI (5 mol%) as a co-catalyst. An extensive study on the mechanism of the reaction was carried out using benzothiazole as a model substrate. After observing that the addition of $^nBu_4NBr$, a known stabilizer of palladium nanoparticles, did not have a beneficial effect on the reaction outcome, and noticing the results of some poisoning assays and of ^{31}P-NMR monitorization, the authors proposed that, in contrast to previous reports on direct arylation reactions, a Pd(II)–Pd(IV)–Pd(II) pathway could be responsible for the presented arylation. As a result, the authors suggested that the catalytic cycle begins with coordination of benzothiazole with CuX to generate copper complex **A**, which turns, after H-2 deprotonation, into species **B**. Alternatively, **B** could be formed by an initial deprotonation followed by interaction with CuI. Copper-benzothiazolyl complex **B** would then promote transmetalation with palladium pincer complex PCN (**POCN**) leading to complex **C**, which was isolated. Oxidation addition of **C** with the aryl iodide would generate octahedral Pd(IV) complex **D** which, upon reductive elimination, would provide the product as well as the initial PCN complex **POCN** (Scheme 5). Some of the suggested intermediates were isolated and submitted to reaction conditions, providing the expected arylated products. Moreover, these results were corroborated by the kinetic studies and DFT calculations performed by the authors [122,123].

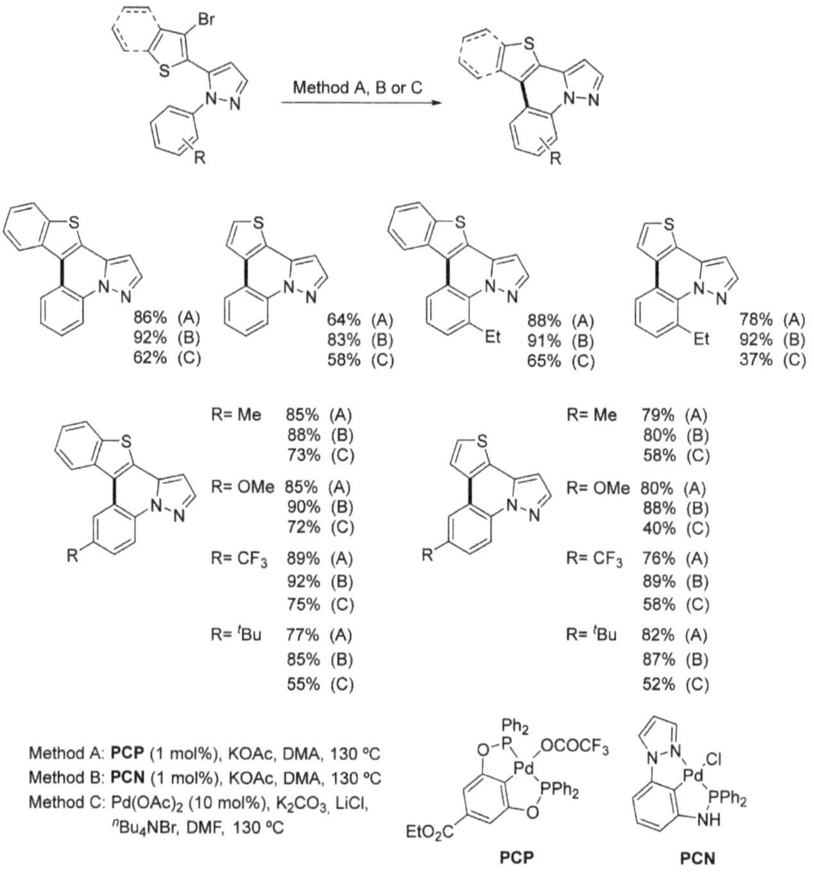

Scheme 4. Synthesis of pyrazolo(benzo)thienoquinolines by direct heteroarylation.

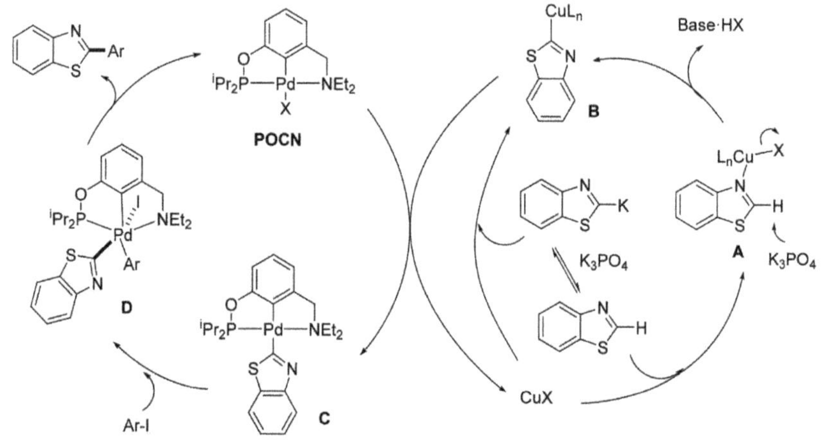

Scheme 5. Mechanistic proposal for the direct heteroarylation of azoles.

As a follow-up research on the results from their previous work on direct heteroarylation [121], in 2015, SanMartin's group reported the intramolecular direct arylation of amides and sulfonamides in the presence of a PCN palladium pincer complex. The addition of a small amount of water was the key for regioselective access to a wide number of structurally diverse phenanthridinone derivatives using an even smaller amount of catalyst (0.05 mol%) of the phosphinoamide-based PCN palladium pincer depicted in Scheme 6. In addition, benzoisothiazoloindole 5,5-dioxides, benzothiazinoindole 7,7-dioxides, benzopyrroloisothiazole 5,5-dioxides and other biologically relevant sulfur heterocycles were prepared from 2-bromobencenesulfonamides under the same reaction conditions (Scheme 3) [124]. A major advantage of the presented method was the low amount of trace palladium impurities in the final products (0.29 ppm, measured by ICP-MS), certainly due to the small catalyst amount employed. The presence of metal traces in final products is a serious concern for the pharmaceutical industry, with increasingly stricter regulations that limit metal contents to 1 ppm or even less depending on the drug administration route (oral, parenteral, nasal, etc.). In many cases, costly and tedious purification steps are required to remove these toxic contaminants [125]. Wang et al. synthesized a phosphinite-based PCN palladium pincer complex comprising a benzimidazole unit and used it for the direct arylation of (benzo)thiazole and benzoxazole with aryl iodides. As in the paper by Punji [122], CuI was added as a co-catalyst. Using cesium carbonate as a base, they were able to reduce the amount of the palladium catalyst to 0.25 mol%. However, in some cases, 0.5 mol% of the pincer complex and 2.5–5 mol% of CuI were required to attain reasonable yields. A Pd(II)–Pd(IV)–Pd(II) pathway was also proposed to explain the role of the above pincer complex in the arylation reaction [126].

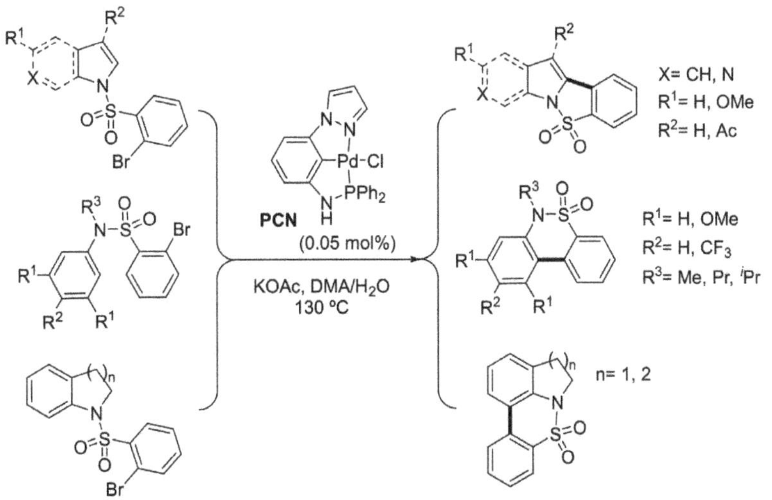

Scheme 6. Intramolecular arylation of *o*-bromobenzenesulfonamides.

3. NHC Containing Pincer Complexes

N-heterocyclic carbenes (NHC) were introduced in organometallic chemistry by Öfele, who reported the first example back in 1968 [127]. Carbene moieties are usually incorporated in bi- and tridentate ligands due to the high stability they can provide to metal complexes [128,129]. Singh's group reported the first use of palladium pincer complexes containing NHC moieties as catalysts for direct arylation reactions. After preparing a series of non-palindromic CNS and CNSe complexes (**C1–C4**) derived from chalcogenated acetamide-functionalized benzimidazolium salts, their catalytic performance in the regioselective C-5 arylation of 1-methyl- and 1,2-dimethylimidazoles with aryl halides under aerobic conditions was examined. A substoichiometric amount of pivalic acid (30 mol%) turned out to be crucial for the reaction outcome. In this regard, the authors proposed that pivalic acid generates coordinatively unsaturated Pd as its proton neutralizes the anionic nitrogen of the amidate fragment, thus helping in the cleavage of the Pd–N bond. That behavior would be consistent with a concerted metalation–deprotonation (CMD) pathway. Benzoic acid was also assayed, providing negligible results. The authors rationalize this outcome considering the higher acidity of benzoic acid, which would stabilize the reaction intermediates so that no further catalysis can occur. As for pincer complexes, 0.5 mol% of catalysts **C1–C4** was chosen as optimal for the reaction scope, illustrated by the arylation with sterically hindered 1-bromonaphthalene, as depicted in Scheme 7a. The regioisomeric identity of some of the arylated products was additionally determined by single-crystal X-ray diffraction analysis. Heteroarylation with 3-bromopyridine and 3-bromoquinoline was also carried out under the same conditions. Minor side-products from the C-4 arylation of imidazole and homocoupling of aryl bromides were also detected. 4-Chlorobenzaldehyde and 4-chlorobenzonitrile were also successfully used as arylating agents, although a higher amount of the catalyst (1 mol%) and longer reaction times (20–24 h) were required, probably due to a more difficult oxidative addition step. In addition, the catalytic life of **C1–C4** was also tested by recycling or reusing these complexes for six runs. Good yields were obtained in all cases, although a steady decrease was observed in every consecutive run [130]. Very similar reaction conditions (K_2CO_3, PivOH, DMA, 110 °C) were used by Joshi and co-workers to effect the direct C-5 arylation of imidazole derivatives with aryl bromides in the presence of an SCSe complex (**C5**), where the NHC moiety occupied the central position of the tridentate ligand (Scheme 7b). Arylation with 4-nitrochlorobenzene was also carried out, although a significant decrease in the reaction yield was observed. In addition, catalyst **C5** demonstrated remarkable recyclability up to five consecutive cycles [131]. Following Finke's report on procedures to distinguish between homogeneous and heterogeneous palladium catalysts [132], both research groups performed poisoning experiments with Hg and PPh_3 (Pd/Hg (1/400), 5 mol % of PPh_3). Palladium nanoparticles and other palladium (0) species amalgamate with mercury so that a noticeable drop in the conversion yield is observed (mercury drop test). However, no inhibition was observed for the above direct arylation reactions after adding overstoichiometric amounts of these poisoning agents. Considering the results from these poisoning assays and the recyclability exhibited by their pincer complexes, the authors suggested that the catalysis was homogeneous in nature [130,131].

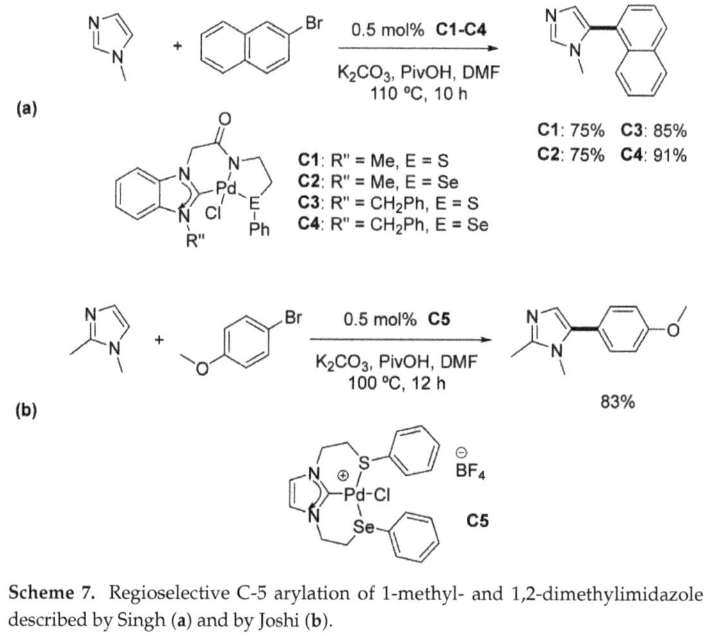

Scheme 7. Regioselective C-5 arylation of 1-methyl- and 1,2-dimethylimidazole derivatives as described by Singh (**a**) and by Joshi (**b**).

The selective arylation of 1,2-dimethylimidazole and imidazo[1,2-*a*]pyridine derivatives with bromoarenes in the presence of 2 mol% of CNO palladium(II) complexes containing NHC moieties (**CNO1–CNO3**) was studied by Lee and co-workers (Scheme 8). They also compared their catalytic activity with that of several palladium sources and ligands (PdCl$_2$, Pd(OAc)$_2$, Pd(OAc)$_2$/PCy$_3$, etc.) and found that their CNO complex was less active than their previously reported Pd(0) complex featuring bidentate NHC and PPh$_2$ moieties [133], which could catalyze arylation with chloroarenes. PEPPSI precatalyst Pd(IPr)(3-ClPy)Cl$_2$ [134] also provided the product from the benchmark reaction, the C-5 arylation of 1,2-dimethylimidazole and 4-bromoacetophenone, with equal efficiency. As in many other reports on direct arylation processes, DMA was the solvent of choice, and a significant decrease in the yields was observed when switching to DMF, THF or toluene. Except for 2-bromobenzonitrile and 2-bromotoluene, all the sterically hindered bromoarenes failed to furnish the desired product from 1,2-dimethylimidazole at 140 °C.

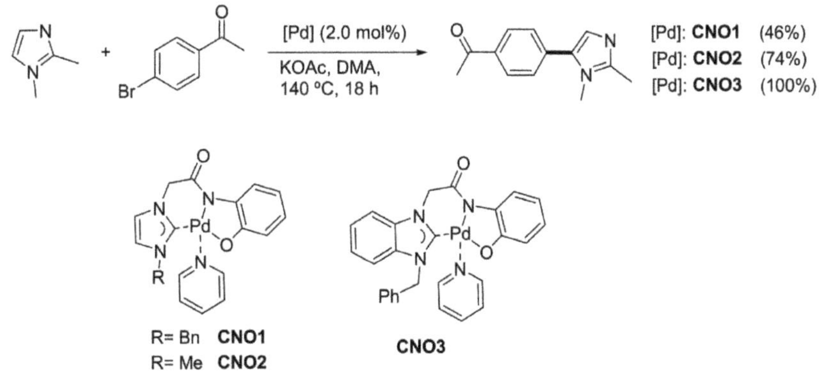

Scheme 8. C-5 arylation of 1,2-dimethylimidazole in the presence of CNO pincer complexes.

In the same paper, imidazo[1,2-*a*]pyridine was reacted with several bromoarenes bearing electron-donating and electron-withdrawing substituents to provide the corresponding 3-arylated compounds with good to excellent yields. In contrast to 1,2-dimethylimidazole, similar yields were achieved for imidazo[1,2-*a*]pyridine when the reaction was carried out under argon and under air. 2-Arylbenzothiophenes were also obtained by reactions between thiophene and arylbromides, and in this case, at lower temperature (110 °C, Scheme 9). Gram-scale reactions (10 mmol) between 1,2-dimethylimidazole and 4-bromobenzonitrile, and between imidazo[1.2-*a*]pyridine and 4-methoxybromobenzene, provided the corresponding arylated compounds in 70% and 66% yields, respectively.

The authors also carried out competitive reactions using an equimolecular mixture of 1,2-dimethylimidazole and imidazopyridine and the same bromoarene. After observing that electron-poor imidazopyridine prevailed over electron-rich 1,2-dimethylimidazole (3-arylimidazopyridine was mainly obtained when using 4-bromoanisole, and exclusively isolated when 4-bromoacetophenone was the arylating agent), they suggested that the arylation proceeds via a Pd(II)–Pd(0)–Pd(II) mechanism based on a concerted metalation–deprotonation (CMD) step (Scheme 10). On account of the electron-donating nature of the 1,2-dimethylimidazole unit and the electron-withdrawing character of the imidazo[1.2-*a*]pyridine core, they synthesized several push–pull chromophores that exhibited a deep blue photoluminescence with moderate quantum efficiency on a large scale, and twisted the intramolecular charge transfer excited state [135].

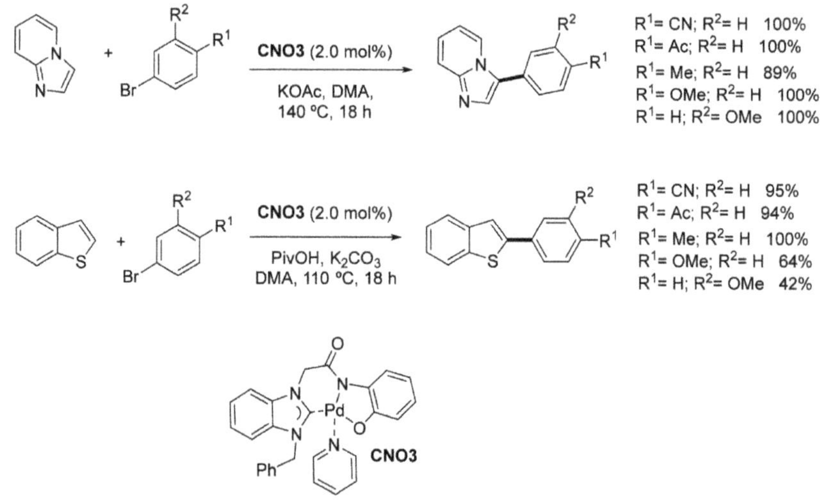

Scheme 9. Direct arylation of imidazo[1,2-*a*]pyridine and benzothiophene.

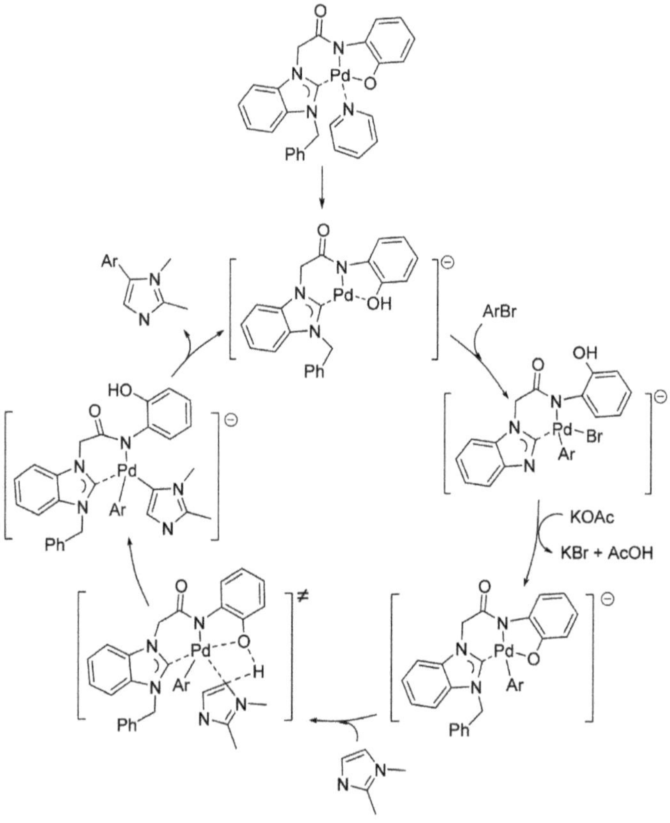

Scheme 10. Mechanism proposed by Lee and co-workers for the C-5 arylation of 1,2-dimethylimidazole.

4. Other Pincer Complexes

Direct arylation has been also reported in the presence of palladium pincer complexes lacking phosphine of NHC moieties. Cai and co-workers prepared a symmetric Schiff-based NCN complex and used it for the selective direct arylation of N-methylindoles at C-2. After some preliminary assays with N-methyl-1H-indole and iodobenzene as model substrates, significantly lower yields (22–46%) were obtained using Pd(OAc)$_2$ or Pd$_2$dba$_3$ (5 mol%) than with their NCN complex (1 mol%). Regarding regioselectivity, reactions carried out in dimethylacetamide at 80 °C provided 2-phenylated product as the only regioisomer, whereas significant amounts of the 3-phenylated product were observed when other solvents (NMP, DMF, AcOH) were used. A further decrease in the catalyst amount to 0.5 mol% diminished the yield to 77%; therefore, the optimized conditions displayed in Scheme 11 (NCN 1 mol%, KOAc, DMA, 80 °C) were applied to a number of aryl iodides, obtaining the corresponding N-methyl-2-arylindoles in moderate to good yields. However, only bromoarenes bearing electron-withdrawing substituents provided acceptable results. Interestingly, 1H-indole and benzothiophene were also regioselectively phenylated with moderate yields under the same conditions. Observation of palladium-black and complete inhibition upon the addition of mercury (mercury drop test) led the authors to propose that their NCN pincer complex is a precatalyst or reservoir of Pd(0) species [136].

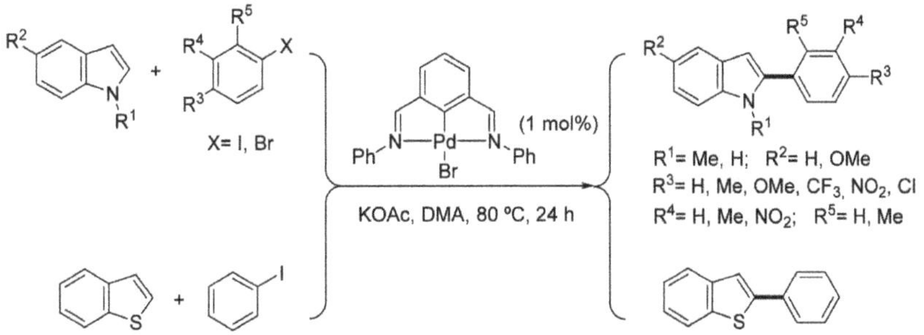

Scheme 11. NCN palladium pincer complex as an active catalyst for the regioselective arylation of indoles.

In addition to their previous work on a PCN complex [122,123], Punji's group prepared several phosphine-free NNN palladium pincer complexes containing a (quinolinyl)amido moiety. One of them turned out to be an efficient catalyst for the direct arylation of benzothiazoles with aryl and heteroaryl iodides in the presence of CuI (1 mol%, Scheme 12). After removal of the arylation product by vacuum distillation and addition of the reagents and solvent, this catalyst was recycled up to five times with a minor decrease in the reaction yield. A hot filtration experiment was performed after the initial heating (30 min, GC yield 34%) to remove all the heterogeneous particles that might account for the slight decrease in the yield observed when adding overstoichiometric amounts of mercury. The reaction was continued upon adding fresh base (K$_2$CO$_3$), and the arylation product was obtained with good yield (88%). On the basis of the results from these two experiments and other mechanistic investigations (kinetic plot, observation of the reactivity order for several aryl iodides, MALDI-TOF-MS analysis of the reaction mixture, etc.), the authors proposed a catalytic cycle akin to that displayed in Scheme 5 [137].

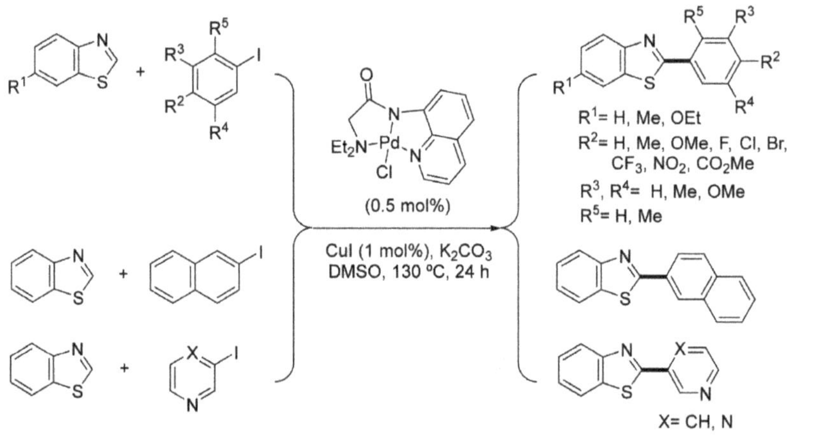

Scheme 12. Direct arylation with aryl iodides in the presence of a NNN palladium pincer complex and CuI.

Maji et al. prepared two ferrocene-based palladium CNN pincer complexes by palladation of the ligands obtained by a condensation of ferrocenecarboxaldehyde with 2-(1-phenylhydrazinyl)pyridine or 2-((1-phenylhydrazinyl)methyl)pyridine. After performing Suzuki–Miyaura biaryl couplings with aryl chlorides, the efficient C-5 arylation of 4-methylthiazole and the C-4 arylation of 3,5-isoxazole with aryl bromides were explored. As for the Suzuki–Miyaura couplings, 0.1 mol% of their CNN complex was enough to catalyze the direct arylation reactions. Good yields were obtained regardless of the electronic nature of the bromoarene. Palladium nanoparticles, generated in situ by decomposition of these pincer complexes, were thought by the authors to be the real catalyst species through a Pd(0)-Pd(II) cycle [138]. A year later, they reported the preparation of four structurally related CNN complexes by the condensation of benzaldehyde derivatives and 2-(1-2-((1-phenylhydrazinyl)methyl)pyridine followed by palladation with Na_2PdCl_4. Optimization of the model reaction, the arylation of 1-methyl-1H-imidazole with 4-bromobenzaldehyde, was carried out with one of the four tricoordinated complexes. Then, the scope of the reaction was expanded and 1-methyl- and 1,2-dimethylimidazole were regioselectively arylated (C-5) with bromoarenes by using 5×10^{-2} mol% of this palladium source (Scheme 13).

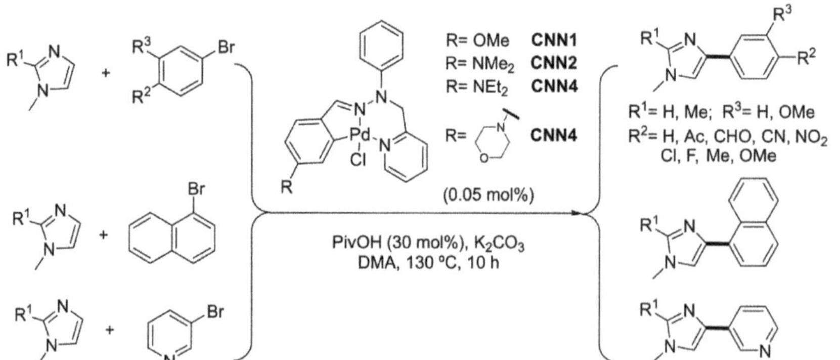

Scheme 13. C-5 arylation with bromoarenes in the presence of **CNN1–CNN4** palladium pincer complexes.

The procedure was also useful for the arylation of the same azoles with aryl chlorides in the presence of another of the four CNN complexes, although a slight increase in the amount of the later palladacycle was required (0.1 mol%). In order to explain the reaction mechanism, the authors proposed the catalytic cycle displayed in Scheme 14. After the in situ generation of palladium(0) species **A**, oxidative addition with the aryl halide provided intermediate **B**, which underwent a ligand exchange with potassium pivalate to form intermediate **C**. Interaction with methylimidazole generated intermediate **E** via a CMD transition state (**D**), and after a reductive elimination step, the arylated product was released along with the Pd(0) species [139].

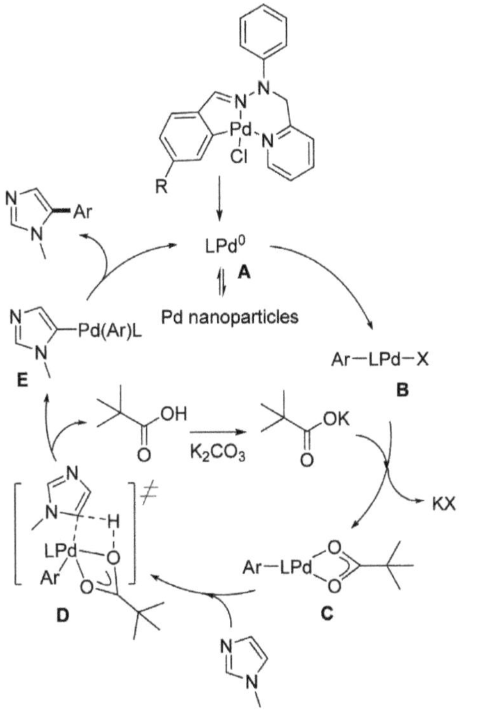

Scheme 14. Possible mechanism for the direct regioselective arylation of 1-methylimidazole derivatives.

Following their research on Heck and Hiyama reactions, Uozumi's group described the use of infinitesimal amounts (5×10^{-3} mol%) of a phenanthroline-based CNN pincer for the direct arylation of (benzo)thiophene derivatives with aryl bromides. Benzothiophenes were arylated with electronically dissimilar and sterically hindered bromoarenes. Regioselective C-5 arylation was observed for 2-substituted and 2,3-disubstituted thiophenes (Scheme 15). Regarding the role of the above CNN complex, palladium nanoparticles (average diameter 3.2 nm) were detected by TEM (transmission electron microscopy) after completion of the reaction. Accordingly, the authors proposed that monomeric palladium(0) species released from the pincer complex were responsible for the catalytic activity observed [140].

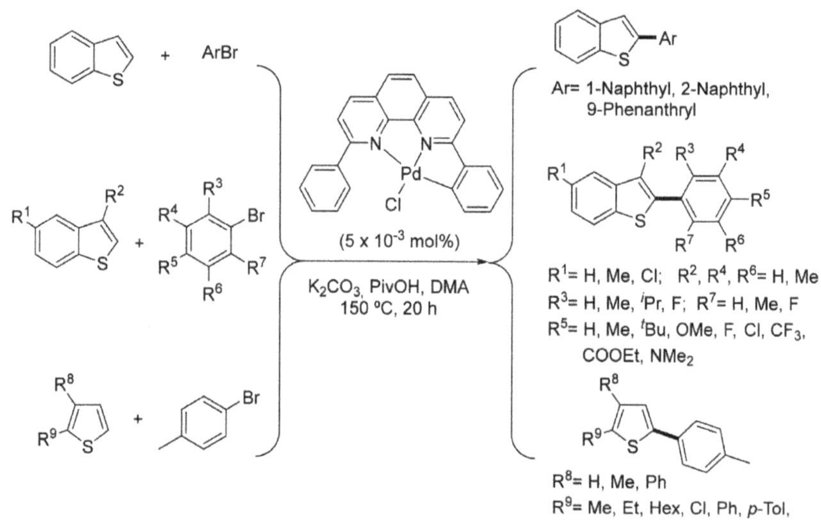

Scheme 15. Direct arylation of thiophene derivatives in the presence of a CNN palladium pincer complex.

5. Conclusions

Direct arylation has been consolidated as an advantageous alternative to cross-coupling reactions involving transmetallating agents. In this regard, the use of palladium pincer complexes as (pre)catalysts for this reaction has attracted much attention because of the lower catalyst amounts required. However, depending on the coupling partners and the complex employed, it is not clear if such efficiency is due to a steady release of palladium(0) species (e.g., palladium nanoparticles) from the complex, to a Pd(II)–Pd(IV) catalytic cycle or to a cocktail of different mechanisms simultaneously taking place. Further research in this field will probably reveal the nature of the true catalysts and will expand the reaction scope by introducing new, more active pincer complexes. Finally, given the almost exclusive use of DMA and DMF as solvents in these reactions, safer reaction media would be also desirable.

Author Contributions: Co-authors M.T.H., N.C. and F.C. contributed to searching and collating the relevant literature and the proof-reading of the document. Co-author G.U. and corresponding author R.S. wrote the body of the article. All authors have read and agreed to the published version of the manuscript.

Funding: This research was funded by the Basque Government (IT1405-19) and the Spanish Ministry of Economy and Competitiveness (CTQ2017-86630-P).

Acknowledgments: Technical and human support provided by SGIker of UPV/EHU is gratefully acknowledged.

Conflicts of Interest: The authors declare no conflict of interest.

References

1. Polascheck, N.; Bankstahl, M.; Löscher, W. The COX-2 inhibitor parecoxib is neuroprotective but not antiepileptogenic in the pilocarpine model of temporal lobe epilepsy. *Exp. Neurol.* **2010**, *224*, 219–233. [CrossRef]
2. Bradley, D.A.; Godfrey, A.G.; Schmid, C.R. Synergistic methodologies for the synthesis of 3-aroyl-2-arylbenzo[b]thiophene-based selective estrogen receptor modulators. Two concise syntheses of raloxifene. *Tetrahedron Lett.* **1999**, *40*, 5155–5159. [CrossRef]
3. Cupido, T.; Rack, P.G.; Firestone, A.J.; Hyman, J.M.; Han, K.; Sinha, S.; Ocasio, C.A.; Chen, J.K. The imidazopyridine derivative JK184 reveals dual roles for microtubules in hedgehog signaling. *Angew. Chem. Int. Ed.* **2009**, *48*, 2321–2324. [CrossRef]
4. Warr, D.; Hesketh, P. Cannabinoids as antiemetics: Everything that's old is new again. *Ann. Oncol.* **2020**, *31*, 1425–1426. [CrossRef]

5. Dale, D.J.; Dunn, P.J.; Golightly, C.; Hughes, M.L.; Levett, P.C.; Pearce, A.K.; Searle, P.M.; Ward, G.; Wood, A.S. The chemical development of the commercial route to sildenafil: A case history. *Org. Proc. Res. Dev.* **2000**, *4*, 17–22. [CrossRef]
6. Larsen, R.D.; King, A.O.; Chen, C.Y.; Corley, E.G.; Foster, B.S.; Roberts, F.E.; Yang, C.; Lieberman, D.R.; Reamer, R.A.; Tschaen, D.M.; et al. Efficient synthesis of Losartan, a nonpeptide angiotensin II receptor antagonist. *J. Org. Chem.* **1994**, *59*, 6391–6394. [CrossRef]
7. Takale, B.S.; Thakore, R.R.; Mallarapu, R.; Gallou, F.; Lipshutz, B.H. A Sustainable 1-pot, 3-step synthesis of Boscalid using part per Million level Pd catalysis in water. *Org. Process Res. Dev.* **2020**, *24*, 101–105. [CrossRef]
8. Li, Z.; Zhang, X.; Qin, J.; Tan, Z.; Han, M.; Jin, G. Efficient and practical synthesis of 3′,4′,5′-trifluoro-[1,1′-biphenyl]-2-amine: A key intermediate of fluxapyroxad. *Org. Process Res. Dev.* **2019**, *23*, 1881–1886. [CrossRef]
9. Szymborski, T.; Cybulski, O.; Bownik, I.; Żywociński, A.; Wieczorek, S.A.; Fiałkowski, M.; Hołyst, R.; Garstecki, P. Dynamic charge separation in a liquid crystalline meniscus. *Soft Matter* **2009**, *5*, 2352–2360. [CrossRef]
10. Dorazco-Gonzalez, A. Chemosensing of chloride based on a luminescent platinum (II) NCN pincer complex in aqueous media. *Organometallics* **2014**, *33*, 868–875. [CrossRef]
11. Wua, Q.; Wang, L. Immobilization of copper (II) in organic-inorganic hybrid materials: A highly efficient and reusable catalyst for the classic Ullmann reaction. *Synthesis* **2008**, *13*, 2007–2012. [CrossRef]
12. Zani, L.; Dessì, A.; Franchi, D.; Calamante, M.; Reginato, G.; Mordini, A. Transition metal-catalyzed cross-coupling methodologies for the engineering of small molecules with applications in organic electronics and photovoltaics. *Coord. Chem. Rev.* **2019**, *392*, 177–236. [CrossRef]
13. Antenucci, A.; Barbero, M.; Dughera, S.; Ghigo, G. Copper catalysed Gomberg-Bachmann-Hey reactions of arenediazonium tetrafluoroborates and heteroarenediazonium o-benzenedisulfonimides. Synthetic and mechanistic aspects. *Tetrahedron* **2020**, *76*, 131632. [CrossRef]
14. Beletskaya, I.P.; Alonso, F.; Tyurin, V. The Suzuki-Miyaura reaction after the nober prize. *Coord. Chem. Rev.* **2019**, *385*, 137–173. [CrossRef]
15. Iffland, L.; Petuker, A.; van Gastel, M.; Apfel, U.-P. Mechanistic implications for the Ni(I)-catalyzed Kumada cross-coupling reaction. *Inorganics* **2017**, *5*, 78. [CrossRef]
16. Lee, V. Application of copper (I) salt and fluoride promoted Stille coupling reactions in the synthesis of bioactive molecules. *Org. Biomol. Chem.* **2019**, *17*, 9095–9123. [CrossRef]
17. Brittain, W.D.G.; Cobb, S.L. Negishi cross-coupling in the synthesis of amino acids. *Org. Biomol. Chem.* **2018**, *16*, 10–20. [CrossRef]
18. Masui, K.; Ikegami, H.; Mori, A. Palladium-catalyzed C−H homocoupling of thiophenes: facile construction of bithiophene structure. *J. Am. Chem. Soc.* **2004**, *126*, 5074–5075. [CrossRef]
19. Seiple, I.B.; Su, S.; Rodriguez, R.A.; Gianatassio, R.; Fujiwara, Y.; Sobel, A.L.; Baran, P.S. Direct C−H arylation of electron-deficient heterocycles with arylboronic acids. *J. Am. Chem. Soc.* **2010**, *132*, 13194–13196. [CrossRef]
20. Yoshikai, N.; Asako, S.; Yamakawa, T.; Ilies, L.; Nakamura, E. Iron-catalyzed C-H bond activation for the ortho-arylation of aryl pyridines and imines with grignard reagents. *Chem. Asian J.* **2011**, *11*, 3059–3065. [CrossRef]
21. Gorelsky, S.I.; Lapointe, D.; Fagnou, K. Analysis of the palladium-catalyzed (aromatic)C−H bond metalation–deprotonation mechanism spanning the entire spectrum of arenes. *J. Org. Chem.* **2012**, *77*, 658–668. [CrossRef]
22. Sandtorv, A.H. Transition metal-catalyzed C-H activation of indoles. *Adv. Synth. Catal.* **2015**, *357*, 2403–2435. [CrossRef]
23. Castro, L.C.M.; Chatani, N. Nichel catalysts/N,N′-bidentate directing groups: An excellent partnership in directed C-H activation reactions. *Chem. Lett.* **2015**, *44*, 410–421. [CrossRef]
24. Simonetti, M.; Perry, G.J.P.; Cambeiro, X.C.; Juliá-Hernández, F.; Arokianather, J.N.; Larrosa, I. Ru-catalyzed C-H arylation of fluoroarenes with aryl halides. *J. Am. Chem. Soc.* **2016**, *138*, 3596–3606. [CrossRef]
25. Pandey, D.K.; Shabade, A.B.; Punji, B. Copper-catalyzed direct arylation of indoles and related (hetero)arenes: A ligandless and solvent-free approach. *Adv. Synth. Catal.* **2020**, *362*, 2534–2540. [CrossRef]
26. Liu, W.; Cao, H.; Lei, A. Iron-catalyzed direct arylation of unactivated arenes with aryl halides. *Angew. Chem. Int. Ed.* **2010**, *49*, 2004–2008. [CrossRef]
27. Hachiya, H.; Hirano, K.; Satoh, T.; Miura, M. Nichel-catalyzed direct arylation of azoles with aryl bromides. *Org. Lett.* **2009**, *11*, 1737–1740. [CrossRef]
28. Join, B.; Yamamoto, T.; Itami, K. Iridium catalysis for C-H bond arylation of heteroarenes with iodoarenes. *Angew. Chem. Int. Ed.* **2009**, *48*, 3644–3647. [CrossRef]
29. Liu, W.; Cao, H.; Xin, J.; Jin, L.; Lei, A. Cobalt-catalyzed direct arylation of unactivated arenes with aryl halides. *Chem. Eur. J.* **2011**, *17*, 3588–3592. [CrossRef]
30. Berman, A.M.; Lewis, J.C.; Bergman, R.G.; Ellman, J.A. Rh (I)-catalyzed direct arylation of pyridines and quinolones. *J. Am. Chem. Soc.* **2008**, *130*, 14926–14927. [CrossRef]
31. Lewis, J.C.; Berman, A.M.; Bergman, R.G.; Ellman, J.A. Rh (I)-catalyzed arylation of heterocycles via C-H bond activation: Expanded scope through mechanistic insight. *J. Am. Chem. Soc.* **2008**, *130*, 2493–2500. [CrossRef]
32. Berman, A.M.; Bergaman, R.G.; Ellman, J.A. Rh (I)-catalyzed direct arylation of azines. *J. Org. Chem.* **2010**, *75*, 7863–7868. [CrossRef]
33. Wang, Q.; Cai, Z.-J.; Liu, C.-X.; Gu, Q.; You, S.-L. Rhodium-catalyzed atroposelective C-H arylation: Efficient synthesis of axially chiral heterobiaryls. *J. Am. Chem. Soc.* **2019**, *141*, 9504–9510. [CrossRef]

34. Ackermann, L.; Vicente, R.; Potukuchi, H.K.; Pirovano, V. Mechanistic insight into direct arylation with ruthenium (II) carboxylate catalysts. *Org. Lett.* **2010**, *12*, 5032–5035. [CrossRef]
35. Al Mamari, H.H.; Diers, E.; Ackermann, L. Triazole-assisted ruthenium-catalyzed C-H arylarion of aromatic amides. *Chem. Eur. J.* **2014**, *20*, 9739–9743. [CrossRef]
36. Roman, D.S.; Poiret, V.; Pelletier, G.; Charette, A.B. Direct arylation of imidazo[1,5-*a*]azines through ruthenium and palladium catalysis. *Eur. J. Org. Chem.* **2015**, *2015*, 67–71. [CrossRef]
37. Zha, G.-F.; Qin, H.-L.; Kantchev, E.A.B. Ruthenium-catalyzed direct arylations with aryl chlorides. *RSC Adv.* **2016**, *6*, 30875–30885. [CrossRef]
38. Kaloğlu, N.; Özdemir, İ.; Gürbüz, N.; Arslan, H.; Dixneuf, P.H. Ruthenium (η^6, η^1-arene-CH$_2$-NHC) catalysts for direct arylation of 2-phenylpyridine with (hetero)aryl chlorides in water. *Molecules* **2018**, *23*, 647. [CrossRef]
39. Lafrance, M.; Rowley, C.N.; Woo, T.K.; Fagnou, K. Catalytic intermolecular direct arylation of perfluorobenzenes. *J. Am. Chem. Soc.* **2006**, *128*, 8754–8756. [CrossRef]
40. Satoh, T.; Miura, M. Catalytic direct arylation of heteroaromatic compounds. *Chem. Lett.* **2005**, *36*, 200–205. [CrossRef]
41. Yuen, O.Y.; Leung, M.P.; Chau Ming So, C.M.; Sun, R.W.-Y.; Kwong, F.Y. Palladium-Catalyzed Direct Arylation of Polyfluoroarenes for Accessing Tetra-*ortho*-Substituted Biaryls: Buchwald-type Ligand Having Complementary −PPh$_2$ Moiety Exhibits Better Efficiency. *J. Org. Chem.* **2018**, *83*, 9008–9017. [CrossRef]
42. Bheeter, C.B.; Bera, J.K.; Doucet, H. Palladium-catalyzed direct arylation of thiophenes bearing SO$_2$R substituents. *J. Org. Chem.* **2011**, *76*, 6407–6413. [CrossRef]
43. Hayashi, S.; Kojima, Y.; Koizumi, T. Highly regioselective Pd/C-catalyzed direct arylation toward thiphene-based-conjugated polymers. *Polym. Chem.* **2015**, *6*, 881–885. [CrossRef]
44. Yokooji, A.; Okazawa, T.; Satoh, T.; Miura, M.; Nomura, M. Palladium-catalyzed direct arylation of thiazoles with aryl bromides. *Tetrahedron Lett.* **2003**, *59*, 5685–5689. [CrossRef]
45. Turner, G.L.; Morris, J.A.; Greaney, M.F. Direct arylation of thiazoles on water. *Angew. Chem. Int. Ed.* **2007**, *46*, 7996–8000. [CrossRef]
46. Ackermann, L.; Vicente, R.; Born, R. Palladium-catalyzed direct arylations of 1,2,3-triazoles with aryl chlorides using conventional heating. *Adv. Synth. Catal.* **2008**, *350*, 741–748. [CrossRef]
47. Ohnmacht, S.A.; Culshaw, A.J.; Greaney, M.F. Direct arylation of 2*H*-Indazoles on water. *Org. Lett.* **2010**, *12*, 224–226. [CrossRef]
48. Ben-Yahia, A.; Naas, M.; El Kazzouli, S.; Essassi, M.; Guillaumet, G. Direct C-3-arylations of 1*H*-Indazoles. *Eur. J. Org. Chem.* **2012**, *2012*, 7075–7081. [CrossRef]
49. Carrër, A.; Brinet, D.; Florent, J.-C.; Rousselle, P.; Bertounesque, E. Palladium-catalyzed direct arylation of polysubstituted benzofurans. *J. Org. Chem.* **2012**, *77*, 1316–1327. [CrossRef]
50. Rampon, D.S.; Wssjohann, L.A.; Schneider, P.H. Palladium-catalyzed direct arylation of selenophene. *J. Org. Chem.* **2014**, *79*, 5987–5992. [CrossRef]
51. Bedford, R.B.; Durrat, S.J.; Montgomery, M. Catalyst-Switchable regiocontrol in the direct arylation of remote C-H groups in pyrazolo[1,5-*a*]pyrimidines. *Angew. Chem. Int. Ed.* **2015**, *54*, 8787–8790. [CrossRef]
52. Kumar, P.V.; Lin, W.-S.; Shen, J.-S.; Nandi, D.; Lee, H.M. Direct C5-arylation reaction between imidazoles and aryl chlorides catalysed by palladium complexes with phosphines and *N*-heterocyclic carbenes. *Organometallics* **2011**, *30*, 5160–5169. [CrossRef]
53. Littke, A.F.; Fu, G.C. Palladium-catalyzed couplikng reactions of aryl chlorides. *Angew. Chem. Int. Ed.* **2002**, *41*, 4176–4211. [CrossRef]
54. Campeau, L.-C.; Thansandote, P.; Fagnou, K. High-yielding intramolecular direct arylation reactions with aryl chlorides. *Org. Lett.* **2005**, *7*, 1857–1860. [CrossRef]
55. Martin, A.R.; Chartoire, A.; Slawin, A.M.Z.; Nolan, S.P. Extending the utility of [Pd(NHC)(cinnamyl)Cl] precatalysts: Direct arylation of heterocycles. *Beilstein J. Org. Chem.* **2012**, *8*, 1637–1643. [CrossRef]
56. Li, Y.; Wang, J.; Huang, M.; Wang, Z.; Wu, Y.; Wu, Y. Direct C-H arylation of thiophenes at low catalyst loading og a phosphine-free bis (alkoxo) palladium complex. *J. Org. Chem.* **2014**, *79*, 2890–2897. [CrossRef]
57. He, X.-X.; Li, Y.; Ma, B.-B.; Ke, Z.; Liu, F-S. Sterically encumbered tetraarylimidazolium carbine Pd-PEPPSI complexes: Highly eddicient direct arylation of imidazoles with aryl bromides under aerobic conditions. *Organometallics* **2016**, *35*, 2655–2663. [CrossRef]
58. Aktaş, A.; Celepci, D.B.; Gök, Y. Nover 2-hydroxyethyl substituted *N*-coordinate-Pd (II) (NHC) and bis (NHC) Pd (II) complexes: Synthesis, characterization and the catalytic activity in the direct arylation reaction. *J. Chem. Sci.* **2019**, *131*, 78. [CrossRef]
59. El Abbouchi, A.; Koubachi, J.; El Brahmi, N.; El Kazzouli, S. Direct arylation and Suzuki-Miyaura coupling of imidazo [1,2-a]pyridines catalysed by (SIPr) Pd (allyl) Cl complex under microwave-irradiation. *Med. J. Chem.* **2019**, *9*, 347–354. [CrossRef]
60. Şhain, N.; Gürbüz, N.; Karbiyik, H.; Karabiyik, H.; Özdemir, İ. Arylation of heterocyclic compounds by benzimidazole-based *N*-heterocyclic carbene-palladium (II) complexes. *J. Organomet. Chem.* **2020**, *907*, 121076. [CrossRef]
61. Kaloğlu, M.; Kaloğlu, N.; Özdemir, I. Palladium-PEPPSI-NHC complexes bearing imidazolidin-2-ylidene ligand: Efficient precatalysts for the direct C5-arylation of *N*-methylpyrrole-2-carboxaldehyde. *Catal. Lett.* **2021**, 1–16. [CrossRef]
62. Sun, H.-Y.; Gorelsky, S.I.; Stuart, D.R.; Campeau, L.-C.; Fagnou, K. Mechanistic analysis of azine *N*-oxide direct arylation: Evidence for a critical role of acetate in the Pd (OAc)$_2$ precatalyst. *J. Org. Chem.* **2010**, *75*, 8180–8189. [CrossRef]

63. Nakao, Y.; Kanyiva, K.S.; Oda, S.; Hiyama, T. Hydroheteroarylation of alkynes under mild nickel catalysis. *J. Am. Chem. Soc.* **2006**, *128*, 8146–8147. [CrossRef]
64. García-Cuadrado, D.; Braga, A.A.C.; Maseras, F.; Echavarren, A.M. Proton abtraction mechanism for the palladium-catalyzed intramolecular arylation. *J. Am. Chem. Soc.* **2006**, *128*, 1066–1067. [CrossRef]
65. Zhang, W.; Zeng, Q.; Zhang, X.; Tian, Y.; Yue, Y.; Guo, Y.; Wang, Z. Ligand-free CuO nanospindle catalysed arylation of heterocycle C-H bonds. *J. Org. Chem.* **2011**, *76*, 4741–4745. [CrossRef]
66. Salamanca, V.; Aléniz, A.C. Faster palladium-catalyzed arylation of simple arenes in the presence of a methylketone: Beneficial effect of an *a piori* interfering solvent in C-H activation. *Org. Chem. Front.* **2021**, *8*, 1941–1951. [CrossRef]
67. Lane, B.S.; Brown, M.A.; Sames, D. Direct palladium-catalyzed C-2 and C-3 arylation of indoles: A mechanistic rationale for regioselectivity. *J. Am. Chem. Soc.* **2005**, *127*, 8050–8057. [CrossRef]
68. Martín-Matute, B.; Mateo, C.; Cárdenas, D.J.; Echavarren, A.M. Intramolecular C-H activation by alkylpalladium(II) complexes: Insights into the mechanism of the palladium-catalyzed arylation reaction. *Chem. Eur. J.* **2001**, *7*, 2341–2348. [CrossRef]
69. Park, C.-H.; Ryabova, V.; Seregin, I.V.; Sromek, A.W.; Gevorgyan, V. Palladium-catalyzed arylation and heteroarylation of indolizines. *Org. Lett.* **2004**, *6*, 1159–1162. [CrossRef]
70. Zolliger, H. Hydrogen isotope effects in aromatic substitution reactions. *Adv. Phys. Org. Chem.* **1964**, *2*, 163–200. [CrossRef]
71. Gallego, D.; Baquero, E.A. Recent advances on mechanistic studies on C–H activation catalyzed by base metals. *Open Chem.* **2018**, *16*, 1001–1058. [CrossRef]
72. Davies, D.L.; Donald, S.M.; Macgregor, S.A. Computational study of the mechanism of cyclometalation by palladium acetate. *J. Am. Chem. Soc.* **2005**, *127*, 13754–13755. [CrossRef]
73. Hennessy, E.J.; Buchwald, S.L. Synthesis of substituted oxindoles from a-chloroacetanilides via palladium-catalyzed C-H functionalization. *J. Am. Chem. Soc.* **2003**, *125*, 12084–12085. [CrossRef]
74. Zhang, C.; Tang, C.; Jiao, N. Recent advances in copper-catalyzed dehydrogenative functionalization via a single electron transfer (SET) process. *Chem. Soc. Rev.* **2012**, *41*, 3464–3484. [CrossRef]
75. Mota, A.J.; Dedieu, A.; Bour, C.; Suffert, J. Cyclocarbopalladation involving an unusual 1,5-palladium vinyl to aryl shift as termination step: Theoretical study of the mechanism. *J. Am. Chem. Soc.* **2005**, *127*, 7171–7182. [CrossRef]
76. Boeglin, D.; Cantel, S.; Heitz, A.; Martinez, J.; Fehrentz, J.-A. Solution and solid-supported synthesis of 3,4,5-trisubstituted 1,2,4-triazole-based peptidomimetics. *Org. Lett.* **2003**, *5*, 4465–4468. [CrossRef]
77. Davies, D.-L.; Macgregor, S.A.; McMullin, C.L. Computational studies of carboxylate-assisted C-H activation and functionalization at group 8–10 transition metal centers. *Chem Rev.* **2017**, *117*, 8649–8709. [CrossRef]
78. Capito, E.; Brown, J.M.; Ricci, A. Directed palladation: Fine tuning permits the catalytic 2-alkenylation of indoles. *Chem. Commun.* **2005**, 1854–1856. [CrossRef]
79. Hughes, C.C.; Trauner, D. Concise total synthesis of (-)-frondosin B using a novel palladium-catalyzed cyclization. *Angew. Chem. Int. Ed.* **2002**, *41*, 1569–1572. [CrossRef]
80. Campo, M.A.; Huang, Q.; Yao, T.; Tian, Q.; Larock, R.C. 1,4-Palladium migration via C-H activation, followed by arylation: Synthesis of fused polycycles. *J. Am. Chem. Soc.* **2003**, *125*, 11506–11507. [CrossRef]
81. Flegeau, E.F.; Bruneau, C.; Dixneuf, P.; Jutand, A. Autocatalysis for C-H bond activation by ruthenium (II) complexes in catalytic arylation of functional arenes. *J. Am. Chem. Soc.* **2011**, *133*, 10161–10170. [CrossRef]
82. Shan, C.; Luo, X.; Qi, X.; Liu, S.; Li, Y.; Yu Lan, Y. Mechanism of Ruthenium-Catalyzed Direct Arylation of C–H Bonds in Aromatic Amides: A Computational Study. *Organometallics* **2016**, *35*, 1440–1445. [CrossRef]
83. Campeau, L.-C.; Bertrand-Laperle, M.; Leclerc, J.-P.; Villemure, E.; Gorelsky, S.; Fagnou, K. C2, C5 and C4 azole N-oxide direct arylation including room-temperature reactions. *J. Am. Chem. Soc.* **2008**, *130*, 3276–3277. [CrossRef]
84. Rousseau, G.; Breit, B. Removable directing groups in organic synthesis and catalysis. *Angew. Chem. Int. Ed.* **2011**, *50*, 2450–2494. [CrossRef]
85. Kim, J.; Jo, M.; So, W.; No, Z. Pd-catalyzed *ortho*-arylation of 3,4-dihydroisoquinolones via C-H bond activation: Synthesis of 8-aryl-1,2,3,4-tetrahydroisoquinolines. *Tetrahedron Lett.* **2009**, *50*, 1229–1235. [CrossRef]
86. Choi, H.; Min, M.; Peng, Q.; Kang, D.; Paton, R.S.; Hong, S. Unraveling innate substrate control in site-selective palladium-catalyzed C–H heterocycle functionalization. *Chem. Sci.* **2016**, *7*, 3900–3909. [CrossRef]
87. Tlahuext-Aca, A.; Lee, S.Y.; Sakamoto, S.; Hartwig, J.F. Direct arylation of simple arenes with aryl bromides by synergistic silver and palladium catalysis. *ACS Catal.* **2021**, *11*, 1430–1434. [CrossRef]
88. Moroni, F.; Cozzi, A.; Chiaguri, A.; Formentini, L.; Camaioni, E.; Pellegrini-Giampietro, D.E.; Chen, Y.; Liang, S.; Zaleska, M.M.; Gonzales, C.; et al. Long-lasting neuroprotection and neurological improvement in stroke models with new, potent and brain permeable inhibitors of poly (ADP-ribose) polymerase. *Br. J. Pharma.* **2012**, *165*, 1487–1500. [CrossRef]
89. Hegan, D.C.; Lu, Y.; Stachelek, G.C.; Crosby, M.E.; Bindra, R.S.; Glazer, P.M. Inhibition of poly (ADP-ribose) polymerase down-regulates BRCA1 and RAD51 in a pathway mediated by E2F4 and p130. *Proc. Natl. Acad. Sci. USA* **2010**, *107*, 2201–2206. [CrossRef]
90. Aoyama, H.; Sugita, K.; Nakamura, M.; Aoyama, A.; Salim, T.A.M.; Okamoto, M.; Baba, M.; Hashimoto, Y. Fused heterocyclic amido compounds as anti-hepatitis C virus agents. *Bioorg. Med. Chem.* **2011**, *19*, 2675–2687. [CrossRef]
91. Dyker, G. *Handbook of C-H Transformations*, 1st ed.; Wiley-VCH: Weinheim, Germany, 2005.

92. Deshpande, P.P.; Martin, O.R. A concise total synthesis of the aglycone of the gilvocarcins. *Tetrahedron Lett.* **1990**, *44*, 6313–6316. [CrossRef]
93. Hemberge, Y.; Zhang, G.; Brun, R.; Kaiser, M.; Bringmann, G. Highly antiplasmodial non-natural oxidative products of dioncophylline A: Synthesis, absolute configuration, and conformational stability. *Chem. Eur. J.* **2015**, *21*, 14507–14518. [CrossRef]
94. Bringmann, G.; Walter, R.; Weirich, R. The Directed synthesis of biaryl compounds: Modern concepts and strategies. *Angew. Chem. Int. Ed.* **1990**, *29*, 977–991. [CrossRef]
95. Bringmann, G.; Pabst, T.; Henschel, P.; Kraus, J.; Peters, K.; Peters, E.-M.; Rycroft, D.S.; Connolly, J.D. Nondynamic and dynamic kinetic resolution of lactones with stereogenic centers and axes: Stereoselective total synthesis of herbertenediol and mastigophorenes A and B. *J. Am. Chem. Soc.* **2000**, *122*, 9127–9133. [CrossRef]
96. Harayama, T.; Yasuda, H.; Akiyama, T.; Takeuchi, Y.; Abe, H. Synthesis of arnottin I through a palladium-mediated aryl-aryl coupling reaction. *Chem. Pharm. Bull.* **2000**, *48*, 861–864. [CrossRef]
97. Hermann, W.A.; Brossmer, C.; Reisinger, C.-P.; Riermeier, T.H.; Öfele, K.; Beller, M. Palladacycles: Efficient new catalyst for the heck vinylation of aryl halides. *Chem. Eur. J.* **1997**, *8*, 1357–1364. [CrossRef]
98. Bringmann, G.; Ochse, M.; Götz, R. First atropo-divergent total synthesis of the antimalarial korupensamines A and B by the "lactone method". *J. Org. Chem.* **2000**, *65*, 2069–2077. [CrossRef]
99. Rice, J.E.; Cai, Z.-W.; He, Z.-M.; LaVoie, E.J. Some observations on the palladium-catalyzed triflate-arene cyclization of electron-rich biaryl substrates. *J. Org. Chem.* **1995**, *24*, 8101–8104. [CrossRef]
100. Wang, L.; Shevlin, P.B. Formation of benzo[*ghi*]fluoranthenes by palladium catalyzed intramolecular coupling. *Tetrahedron Lett.* **2000**, *41*, 285–288. [CrossRef]
101. Echavarren, A.M.; Gómez-Lor, B.; González, J.J.; de Frutos, O. Palladium-catalyzed intramolecular arylation reaction: Mechanism and application for the synthesis of polyarenes. *Synlett* **2003**, *2003*, 585–597. [CrossRef]
102. Moulton, C.J.; Shaw, B.L. Transition metal-carbon bonds. Part XLII. Complexes of nickel, palladium, platinum, rhodium and iridium with the tridentate ligand 2,6-bis[(di-t-butylphosphino)methyl]phenyl. *J. Chem. Soc. Dalton Trans.* **1976**, 1020–1024. [CrossRef]
103. Morales-Morales, D.; Jensen, C.M. *The Chemistry of Pincer Compounds*, 1st ed.; Elsevier: Amsterdam, The Netherlands, 2007.
104. Szabó, K.J.; Wendt, O.F. *Pincer and Pincer-Type Complexes*, 1st ed.; Wiley-VCH: Weinheim, Germany, 2014.
105. Singleton, J.T. The use of pincer complexes in organic synthesis. *Tetrahedron* **2003**, *59*, 1837–1857. [CrossRef]
106. Morales-Morales, D.; Redón, R.; Yung, C.; Jensen, C.M. Dehydrogenation of alkanes catalyzed by an iridium phosphinito PCP pincer complex. *Inorg. Chimica Acta* **2004**, *357*, 2953–2956. [CrossRef]
107. Hao, X.-Q.; Wang, Y.-N.; Liu, J.-R.; Wang, K.-L.; Gong, J.-F.; Song, M.-P. Unsymmetrical, oxazolinyl-containing achiral and chiral NCN pincer ligand precursors and their complexes with palladium (II). *J. Organomet. Chem.* **2010**, *695*, 82–89. [CrossRef]
108. Liu, N.; Li, X.; Sun, H. Synthesis and properties of novel ortho-metalated cobalt (I) and iron (II) complexes through C_{sp2}-H bond activation of dibenzylphenylphosphine. *J. Organomet. Chem.* **2011**, *696*, 2537–2542. [CrossRef]
109. Gunanathan, C.; Milstein, D. Bond activation and catalysis by ruthenium pincer complexes. *Chem. Rev.* **2014**, *114*, 12024–12087. [CrossRef]
110. Shih, W.-C.; Gu, W.; Macinnis, M.C.; Herbert, D.E.; Ozerov, O.V. Bory/borane interconversion and diversity of binding modes of oxygenous ligands in PBP pincer complexes of rhodium. *Organometallics* **2017**, *36*, 1718–1726. [CrossRef]
111. Mukherjee, A.; Milstein, D. Homogeneous catalysis by cobalt and manganese pincer complexes. *ACS Catal.* **2018**, *8*, 11435–11469. [CrossRef]
112. Churruca, F.; SanMartin, R.; Tellitu, I.; Domínguez, E. N-heterocyclic NCN-pincer palladium complexes: A source for general, highly efficient catalysts in Heck, Suzuki, and Sonogashira coupling reactions. *Synlett* **2005**, *2005*, 3116–3120. [CrossRef]
113. Takenaka, K.; Uozumi, Y. Development of chiral pincer palladium complexes bearing a pyrroloimidazolone unit. Catalytic use for asymmetric Michael addition. *Org. Lett.* **2004**, *6*, 1833–1835. [CrossRef]
114. Sun, Y.; Koehler, C.; Tan, R.; Annibale, V.T.; Song, D. Ester hydrogenation catalyzed by Ru-CNN pincer complexes. *Chem. Commun.* **2011**, *47*, 8349–8351. [CrossRef]
115. Serra, D.; Cao, P.; Cabrera, J.; Padilla, R.; Rominger, F.; Limbach, M. Development of platinum(I) and –(IV) CNC pincer complexes and their application in a hydrovinylation reaction. *Organometallics* **2011**, *30*, 1885–1895. [CrossRef]
116. Moure, M.J.; SanMartin, R.; Domínguez, E. Copper pincer complexes as advantageous catalysts for the heteroannulation of *ortho*-halophenols and alkynes. *Adv. Synth. Catal.* **2011**, *356*, 2070–2080. [CrossRef]
117. Kim, D.; Le, L.; Drance, M.J.; Jensen, K.H.; Bofdanovski, K.; Cervarich, T.N.; Barnard, M.G.; Pudalov, N.J.; Knapp, S.M.M.; Chianese, A.R. Ester hydrogenation catalysed by CNN-pincer complexes of ruthenium. *Organometallics* **2016**, *35*, 982–989. [CrossRef]
118. Urgoitia, G.; SanMartin, R.; Herrero, M.T.; Domínguez, E. Efficient copper-free aerobic alkyne homocoupling in polyethylene glycol. *Environ. Chem. Lett.* **2017**, *15*, 157–164. [CrossRef]
119. Gorgas, N.; Alves, L.G.; Stöger, B.; Martins, A.M.; Veiros, L.F.; Kirchner, K. Stable, yet highly reactive nonclassical iraon(II) polyhydide pincer complexes: Z-selective dimerization and hydroboration of terminal alkynes. *J. Am. Chem. Soc.* **2017**, *139*, 8130–8133. [CrossRef]
120. González-Sebastián, L.; Morales-Morales, D. Cross-coupling reactions catalysed by palladium pincer complexes. A review of recent advances. *J. Organomet. Chem.* **2019**, *893*, 39–51. [CrossRef]

121. Churruca, F.; Hernández, S.; Perea, M.; SanMartin, R.; Domínguez, E. Direct Access to pyrazolo(benzo)thienoquinolines. Highly effective palladium catalysts for the intramolecular C-H heteroarylation of arenes. *Chem. Commun.* **2013**, *49*, 1413–1415. [CrossRef]
122. Khake, S.M.; Soni, V.; Gonnade, R.G.; Punji, B. Design and development of POCN-pincer palladium catalysts for C-H bond arylation of azoles with aryl iodides. *Dalton Trans.* **2014**, *43*, 16084–16096. [CrossRef]
123. Khake, S.M.; Jagtap, R.A.; Dangat, Y.B.; Gonnade, R.G.; Vanka, K.; Punji, B. Mechanistic insights into pincer-ligated palladium-catalyzed arylation of azoles with aryl iodides: Evidence of a Pd^{II}-Pd^{IV}-Pd^{II} pathway. *Organometallics* **2016**, *35*, 875–886. [CrossRef]
124. Conde, N.; Churruca, F.; SanMartin, R.; Herrero, M.T.; Domínguez, E. A further decrease in the catalyst loading for the palladium-catalyzed direct intramolecular arylation of amides and sulphonamides. *Adv. Synth. Catal.* **2015**, *357*, 1525–1531. [CrossRef]
125. Benaglia, M. *Recoverable and Recyclable Catalysts*; John Wiley & Sons: Chippenham, UK, 2009.
126. Wang, C.; Li, Y.; Lu, B.; Hao, X.-Q.; Gong, J.-F.; Song, M.-P. (Phosphinito)aryl benzimidazole PCN pincer palladium (II) complexes: Synthesis, characterization and catalytic activity in C-H arylation of azoles with aryl iodides. *Polyhedron* **2018**, *143*, 184–192. [CrossRef]
127. Öfele, K. 1,3-dimethyl-4-imidazolinyliden-(2)-pentacarbonylchrom ein neuer übergangsmetall-carben-komplex. *J. Organomet. Chem.* **1968**, *12*, P42–P43. [CrossRef]
128. Lee, J.-Y.; Lee, J.-Y.; Chang, Y.-Y.; Hu, C.-H.; Wang, N.M.; Lee, H.M. Palladium complexes with tridentate N-heterocyclic carbine ligands: Selective "normal" and "abnormal" binding and thir anticancer activities. *Organometallics* **2015**, *34*, 4359–4368. [CrossRef]
129. Corberán, R.; Mas-Marzá, E.; Peris, E. Mono-, bi- and tridentate N-heterocyclic carbene ligands for the preparation of transition-metal-based homogeneous catalysts. *Eur. J. Inorg. Chem.* **2009**, *2009*, 1700–1716. [CrossRef]
130. Bhaskar, R.; Sharma, A.K.; Singh, A.K. Palladium (II) complexes of N-heterocyclic carbine amidates derived from chalcogenated acetamide-functionalized 1H-benzimidazolium salts: Recyclable catalyst for regioselective arylation of imidazoles under aerobic conditions. *Organometallics* **2018**, *37*, 2669–2681. [CrossRef]
131. Bhatt, R.; Bhuvanesh, N.; Sharma, K.N.; Joshi, H. Palladium complexes of thio/seleno-ether containing N-heterocyclic carbenes: Efficient and reusable catalyst for regioselective C-H bond arylation. *Eur. J. Inorg. Chem.* **2020**, *2020*, 532–540. [CrossRef]
132. Widegren, J.A.; Finke, R.G. A review of the problem of distinguishing true homogeneous catalysis from soluble or other metal-particle heterogeneous catalysis under reducing conditions. *J. Mol. Catal. A Chem.* **2003**, *198*, 317–341. [CrossRef]
133. Lee, J.-Y.; Shen, J.-S.; Tzeng, R.-J.; Lu, I.-C.; Lii, J.-H.; Hu, C.-H.; Lee, H.M. Well-defined palladium (0) complexes bearing N-heterocyclic carbene and phosphine moieties: Efficient catalytic applications in Mizoroki-Heck reaction and direct C-H fuctionalization. *Dalton Trans.* **2016**, *45*, 10375–10388. [CrossRef]
134. Chartoire, A.; Frogneux, X.; Boreux, A.; Slawin, A.M.Z.; Nolan, S.P. [Pd(IPr*)(3-Cl-pyridinyl)Cl$_2$]: A novel and efficient PEPPSI precatalyst. *Organometallics* **2012**, *31*, 6947–6951. [CrossRef]
135. Li, H.-H.; Maitra, R.; Kuo, Y.-T.; Chen, J.-H.; Hu, C.-H.; Lee, H.M. A tridentate CNO-donor palladium (II) complex as a efficient catalyst for direct C-H arylation: Application in preparation of imidazole-based push-pull chromophores. *Appl. Organomet. Chem.* **2018**, *32*, 3956. [CrossRef]
136. Feng, J.; Lu, G.; Lv, M.; Cai, C. Palladium catalyzed direct C-2 arylation of indoles. *J. Organomet. Chem.* **2014**, *761*, 28–31. [CrossRef]
137. Pandiri, H.; Soni, V.; Gonnade, R.G.; Punji, B. Development of (quinolinyl)amino-based pincer palladium complexes: A robust and phosphine free catalyst system for C-H arylation of benzothiazoles. *New J. Chem.* **2017**, *41*, 3543–3554. [CrossRef]
138. Maji, A.; Singh, A.; Mohanty, A.; Maji, P.K.; Ghosh, K. Ferrocenyl palladacycles derived from unsymmetric pincer-type lignads: Evidence of Pd (0) nanoparticle generation during Suzuki-Miyaura reaction and applications in the direct arylation of thiazoles and isoxazoles. *Dalton Trans.* **2019**, *48*, 17083–17096. [CrossRef]
139. Maji, A.; Singh, O.; Singh, S.; Mohanty, A.; Maji, P.K.; Ghosh, K. Palladium-based catalysts supported by unsymmetric XYC-1 type pincer ligands: C5 arylation of imidazoles and synthesis of octinoxate utilizing the Mizoroki-Heck reaction. *Eur. J. Inorg. Chem.* **2020**, 1596–1611. [CrossRef]
140. Purta, A.E.; Ichii, S.; Tazawa, A.; Uozumi, Y. C-H arylation of thiophenes with aryl bromides by a parts-per-million loading of a palladium NNC-pincer complex. *Synlett* **2020**, *31*, 1634–1638. [CrossRef]

Review

Rhodium-Catalyzed Synthesis of Organosulfur Compounds Involving S-S Bond Cleavage of Disulfides and Sulfur

Mieko Arisawa * and **Masahiko Yamaguchi**

Department of Organic Chemistry, Graduate School of Pharmaceutical Sciences, Tohoku University, Aoba, Sendai 980-8578, Japan; yama@m.tohoku.ac.jp
* Correspondence: mieko.arisawa.d2@tohoku.ac.jp; Tel.: +81-22-795-6814

Academic Editor: Bartolo Gabriele
Received: 17 July 2020; Accepted: 5 August 2020; Published: 7 August 2020

Abstract: Organosulfur compounds are widely used for the manufacture of drugs and materials, and their synthesis in general conventionally employs nucleophilic substitution reactions of thiolate anions formed from thiols and bases. To synthesize advanced functional organosulfur compounds, development of novel synthetic methods is an important task. We have been studying the synthesis of organosulfur compounds by transition-metal catalysis using disulfides and sulfur, which are easier to handle and less odiferous than thiols. In this article, we describe our development that rhodium complexes efficiently catalyze the cleavage of S-S bonds and transfer organothio groups to organic compounds, which provide diverse organosulfur compounds. The synthesis does not require use of bases or organometallic reagents; furthermore, it is reversible, involving chemical equilibria and interconversion reactions.

Keywords: rhodium; catalysis; synthesis; organosulfur compounds; S-S bond cleavage; chemical equilibrium; reversible reaction

1. Introduction

1.1. Structure and Reactivity of Organic Disulfides

Organosulfur compounds containing C-S bonds are widely used for the manufacture of drugs and materials. Compared with organic compounds containing oxygen, which is another group 16(6A) element, different properties appear owing to the large size and polarizability of sulfur atoms. A characteristic feature of inorganic and organic sulfur compounds is the involvement of different oxidation states (between −2 and +6) of sulfur atoms, which give rise to diverse sulfur functional groups [1]. Thiols (RSH) and sulfonic acids (RSO_3H) are organosulfur compounds with low and high oxidation states, respectively. Sulfenic acids (RSOH) and sulfinic acids (RSO_2H) exhibit intermediate oxidation states. These sulfur acids can form ester and amide derivatives. Elemental sulfur in the oxidation state of 0 is a convenient source of organosulfur compounds.

Among organic functional groups containing sulfur, disulfides (RS-SR) with S-S bonds are of interest. In contrast to peroxides (RO-OR) with O-O bonds, disulfides are stable and exhibit significantly different reactivities. The bond energy of S-S bonds is 226 kJ mol^{-1} (for S_8) [2–5], which is the highest among the X-X bonds of the group 16 elements: O-O, 142 kJ mol^{-1}; Se-Se, 172 kJ mol^{-1}; and Te-Te, 150 kJ mol^{-1}. Disulfides have a molecular structure with a dihedral angle of C-S-S-C of approximately 90° in the most stable conformation.

Proteins and peptides contain disulfides that form their three-dimensional structures [6–9] Disulfides in proteins are found predominantly in secreted extracellular proteins, and thiols are preserved in the cytosol in a reductive environment. It is known that the functions of proteins can be switched via the cleavage or formation of disulfides is known [10,11]. Disulfides are found in some small molecules that are biologically active natural products [12].

Sulfides, disulfides, and polysulfides are important functional groups in synthetic rubber [13]. Natural rubber is treated with sulfur to convert it into materials with a large range of hardness, elasticity, and mechanical durability, in which sulfur atoms form cross-linking bridges between polymer chains in a process called vulcanization.

An important reversible reaction of disulfides is their reduction to thiols, which may be oxidized to disulfides (Figure 1). Such reactivity has been utilized in biological systems and also in synthetic systems for molecular switching [14]. Various reactions have been reported for the interconversion. The exchange reaction of S-S bonds between disulfides is another important reversible reaction (Scheme 1) [15,16]. The reaction of two different disulfides produces a statistical 1:1:2 mixture of two symmetric disulfides and an unsymmetric disulfide under chemical equilibrium when their thermodynamic stabilities are comparable.

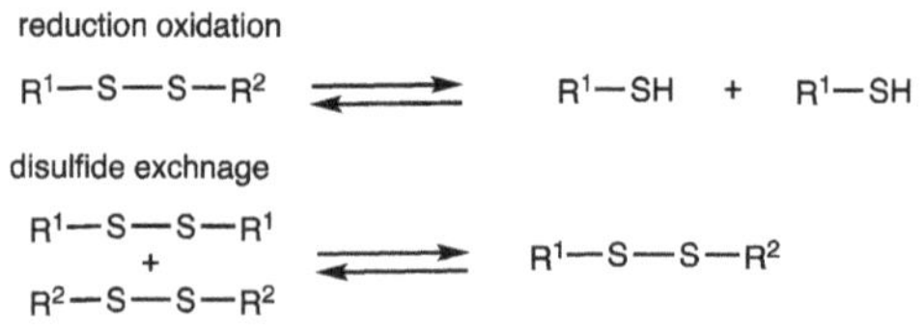

reduction oxidation
$R^1-S-S-R^2 \rightleftharpoons R^1-SH + R^1-SH$

disulfide exchnage
$R^1-S-S-R^1 + R^2-S-S-R^2 \rightleftharpoons R^1-S-S-R^2$

Figure 1. Interconversion reactions of disulfides/thiols by reduction/oxidation and chemical equilibrium under disulfide exchange.

Both reduction/oxidation and exchange reactions of disulfides can be used for a molecular switching function. A characteristic aspect of disulfide exchange reactions compared with reduction/oxidation reactions is that direct one-step interconversion proceeds without forming thiols. This makes procedures simple, and various transformations that are incompatible in the presence of thiols can be conducted. Basic conditions are often employed for disulfide exchange reactions, which involve the nucleophilic attack of thiolate anions on S-S bonds via an S_N2 mechanism. Photoirradiation is effective for the exchange reaction of aromatic disulfides, which involves the homolytic cleavage of S-S bonds generating thiyl radicals. Acidic conditions can also be employed. The reactivity of disulfides depends on their substituents. Aromatic disulfides are easier to dissociate than aliphatic disulfides, which reflects the relative dissociation energies of PhS-SPh (230 kJ mol^{-1}) and MeS-SMe (309 kJ mol^{-1}) [3].

1.2. Synthesis of Organosulfur Compounds Using Disulfides

Synthesis of organosulfur compounds has generally been conducted using thiols, and a typical method is a substitution reaction with organohalogen compounds in the presence of a base [17–20]. The roles of bases are to form highly reactive thiolate anions and to neutralize hydrogen halides formed as byproducts. The neutralization reaction is significantly exothermic and promotes the reaction according to the Bell–Evans–Polanyi principle [17].

The use of disulfides in the substitution reaction of organohalogen compounds has been rare. This is because disulfides are neutral compounds and are less reactive than thiolate anions. Consider a hypothetical substitution reaction of an organohalogen compound and a disulfide to provide a sulfide containing a C-S bond. Formally, the reaction is accompanied by the formation of a sulfenyl halide, which is thermodynamically unstable and makes the reaction thermodynamically unfavorable. The use of disulfides in organic synthesis, however, can have several advantages over the use of thiolate anions: (1) disulfides are stable and easy to handle; (2) they are less odiferous; (3) they can be activated by various methods, including the use of acids, bases, radicals, metals, and photoirradiation; and (4) they do not form metal halide byproducts. In addition, characteristic reactivities of disulfides can appear, which thiols do not have. As such, special methods are needed to utilize disulfides in the synthesis of organosulfur compounds. An example was reported in the oxidation–reduction condensation of bi(2-pyridyl) disulfide and a carboxylic acid in the presence of triphenylphosphine to provide a 2-pyridylthio ester [21]. The reaction is thermodynamically favorable because of the exothermic oxidation reaction of triphenylphosphine to the corresponding oxide. Thus, it is reasonable to consider that disulfides can be used as substrates in organic synthesis and that the reactions can proceed in the absence of bases.

1.3. Rhodium-Catalyzed Synthesis of Organosulfur Compounds Using Disulfides

Transition-metal catalysis for the synthesis of organosulfur compounds has attracted interest; however, this interest has been limited. In particular, the use of disulfides has been rare, with the only exceptions of addition reactions to alkenes and alkynes originally reported by Ogawa [22], Beletskaya [23], and Mitsudo [24]. The lack of such synthetic methods for organosulfur compounds is due to the strong bonding between transition metals and sulfur atoms, which does not readily allow the liberation of the metals and products; as a result, the catalyst cannot be regenerated. Methods are required to overcome the relatively unreactive nature of neutral disulfides and to prevent catalyst deactivation by (1) the development of highly active catalysts and (2) the judicious choice of substrates and products that produce exothermic reactions.

We have found that rhodium complexes catalyze various substitution and insertion reactions using disulfides, which indicates that sulfur ligands on rhodium atoms liberate organosulfur compounds with regeneration of the rhodium catalyst. In this article, a substitution reaction is defined as the transformation of a S-S bond and an X-Y single bond to form a S-X bond; an insertion reaction is defined as the transformation of a S-S bond and an X=Y multiple bond to form a S-X-S subunit with one atom of X inserted or a S-X-Y-S subunit with two atoms of X-Y inserted (Figure 2). We describe in this article that rhodium-catalyzed activation of S-S bonds can be applied to a broad range of chemical transformations with different organic compounds containing S-S, Se-Se, Te-Te, P-P, and P-S heteroatom bonds, along with C-S, C-P, C-F, C-N, C-O, C-H, C-C, and H-H bonds. Unsaturated C=O, C≡C, and C=N bonds can also participate in the rhodium-catalyzed reactions of S-S bonds.

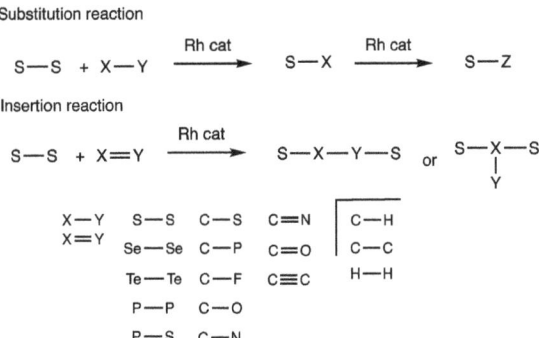

Figure 2. Diverse reactivity of substitution and insertion reactions of the S-S bond in disulfides with various other chemical bonds.

A characteristic feature of rhodium catalysis is its capability to activate C-H bonds [25], which is utilized here for C-S bond formation of 1-alkynes, nitroalkanes, malonates, α-phenylketones, ketones, aldehydes, and heteroaromatic compounds. Activation of C-C bonds in ketones and H-H bonds in hydrogen is also shown.

Another notable property of rhodium-catalyzed synthesis using disulfides is its applicability to reactions in water. Peptides and proteins containing the cysteine moiety can be modified in water without protecting groups (Schemes 2, 3, and 16). These results indicate a high tolerance of functional groups in rhodium catalysis, which selectively activates disulfides in the presence of a large number of oxygen- and nitrogen-containing functional groups.

The most stable form of elemental sulfur is the eight-membered S_8 ring containing S-S bonds, which is produced by desulfurization of petroleum. Sulfur exhibits chemical reactivities similar to those of disulfides, but different chemical reactivities also appear. This is because disulfides contain sulfur atoms bonded to one sulfur atom and one organic group; sulfur contains sulfur atoms bonded to two sulfur atoms. Although organic synthesis using sulfur has attracted attention [26], it has generally been conducted by thermal reactions involving sulfur radicals and by nucleophilic reactions using highly reactive main group metal reagents. In this study, synthesis of organosulfur compounds using disulfides under rhodium catalysis is extended to syntheses using sulfur.

Along with the development of rhodium-catalyzed synthesis of organosulfur compounds involving S-S bond cleavage, we have studied the synthesis of organophosphorus compounds involving P-P bond cleavage. This comparative study of organoheteroatom compounds with elements adjacent on the periodic table is a novel approach.

1.4. Reversible Nature of Rhodium-Catalyzed Reactions of Disulfides

Transition-metal-catalyzed reactions using disulfides to provide organosulfur compounds often reach chemical equilibria and are reversible. This behavior implies that the relative thermodynamic stabilities of substrates and products are similar and the energy barrier is low (Figure 3a). Consequently, shifting the chemical equilibrium toward the desired product is critical to obtaining high yields. In this study, we developed several methods for that purpose: (1) structures of substrates and products are selected to provide exothermic reactions; (2) chemical equilibrium is shifted using a larger amount of one substrate; (3) chemical equilibrium is shifted, removing a volatile product; (4) appropriate combinations of cosubstrates and coproducts are developed to provide exothermic reactions; (5) a product is converted to thermodynamically stable form; and (6) the desired product is removed from an equilibrium mixture by silica nanoparticle precipitation. In this way, the nature of the synthetic reactions for organosulfur compounds, presented herein, is significantly different from that of conventional irreversible reactions using thiolate anions, which involve strong exothermic reactions using bases (Figure 3b).

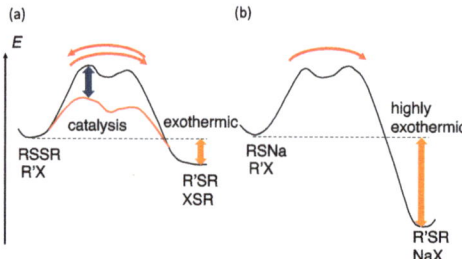

Figure 3. (**a**) Chemical equilibrium of rhodium-catalyzed reactions using disulfides and a cosubstrate; (**b**) highly exothermic irreversible reaction using metal thiolates.

It is thought that chemical equilibrium is not favorable in organic synthesis because chemical yields are governed by the relative thermodynamic stability of substrates and products. We show in this article that the use of chemical equilibrium has intrinsic synthetic advantages: (1) chemical equilibrium is energy-saving because it does not require a strong exothermic reaction and it involves a small energy barrier; (2) chemical equilibrium can provide different products by shifting the equilibrium through the control of reaction conditions; (3) regeneration of substrates from products is easy; (4) catalysis is effective for both forward and backward reactions, which can be used to promote the reaction;

(5) chemical equilibria generally do not form inorganic byproducts, which is inevitable in exothermic reactions using bases. It should also be noted that the recovery of bases from inorganic byproducts is tedious and energy-consuming.

A novel concept is derived from the reversible nature of the rhodium-catalyzed synthesis of organosulfur compounds under chemical equilibrium. Catalysts can cleave C-S bonds of products, which implies that, in the presence of suitable acceptors, products can be converted to other organosulfur compounds: rhodium complexes can activate S-X bonds in the products and convert then into compounds with S-Z bonds (Figure 2). Such examples are described for the reactions of 1-thioalkynes and thioesters in Section 7. Our working hypothesis on the mechanisms of rhodium-catalyzed reactions of disulfides is the involvement of oxidative addition of low-valent rhodium to form S-Rh-S species, which is followed by substitution or insertion by other organic groups. Chemical equilibrium indicates that all processes are reversible.

Organosulfur compounds can be used as alkylating reagents analogous to organohalogen compounds. This concept is inferred from the bond energy of C-S (272 kJ mol^{-1}), which is comparable to that of C-Br (285 kJ mol^{-1}) [1]. Organothio groups can be used as leaving groups in substitution reactions, and they can exhibit different properties from halogen groups. Sulfur leaving groups have divalent sulfur atoms, whereas halogen leaving groups have monovalent atoms, and the properties of the sulfur leaving groups can be tuned by the organic groups. In addition, substitution reactions using sulfur leaving groups are reversible under rhodium catalysis; such examples are described in Section 7. It should be noted that the bond energy of C-S is comparable to that of C-P (264 kJ mol−1) and that organosulfur and organophosphorus compounds are interconvertible (Schemes 14, 18, and 33).

The above describes the reversibility of reactions under chemical equilibrium between **X** and **Y** in a closed system, in which no matter is exchanged with the surroundings but energy is exchanged (Figure 4a). Such reactions are indicated in this article by two straight arrows pointing in opposite directions. A reversible reaction in an open system can also be considered, in which both matter and energy are exchanged. Substrate **X** and product **Y** are interconverted by the addition of reagents **A** and **C**, which provide the **X** + **A** → **Y** + **B** and **Y** + **C** → **X** + **D** reactions, respectively (Figure 4b). Addition of **A** and **C** makes these reactions exothermic, and catalysis reduces the energy barrier, which accelerates these reactions. Such reactions are called interconversion reactions in this article and are expressed by curved arrows pointing in opposite directions. Many of the reactions described in this article are reversible and involve catalysis either under chemical equilibria or interconversion reactions. Reversibility in synthetic reactions provides a novel concept in chemistry that has some similarity with biological systems.

The model of interconversion reactions in an open system between **X** and **Y** provides another interesting aspect in the development of synthetic reactions (Figure 4b). When the model is analyzed in terms of input and output, the **A** + **C** → **B** + **D** reaction can be considered. Such reactions can be developed, as shown in this article (Schemes 5, 7, and 14). **X** and **Y** can be used as catalysts for the **A** + **X** → **B** + **Y** and **Y** + **C** → **X** + **D** reactions proceeding under the same conditions.

One of our purposes in the study of transition-metal-catalyzed synthesis of organosulfur compounds is the development of biologically active compounds. Heteroatoms are essential for biological functions, as shown by the huge amounts of nitrogen and oxygen atoms used by living things. In contrast, the use of sulfur is limited mostly to amino acids, and the development of biologically active organosulfur compounds, which exhibit exotic properties for living things, is an interesting subject. The transition-metal-catalyzed synthetic method discussed herein has indeed provided biologically active organosulfur compounds [26,27].

Our previous review articles on the transition-metal-catalyzed synthesis of organosulfur compounds focused on the development of exothermic reactions [28,29], the synthesis and properties of bis(heteroaryl) compounds [26,27], the use of elemental sulfur [30], and the P-P bond cleavage reactions [31]. The present article describes an overview of our studies that began in the early 2000s and involve classification of the chemical reactions with S-S bond cleavage and organothio transfer.

In addition, emphasized herein are the chemical equilibria and interconversion reactions provided by the reversible nature of the rhodium-catalyzed synthesis.

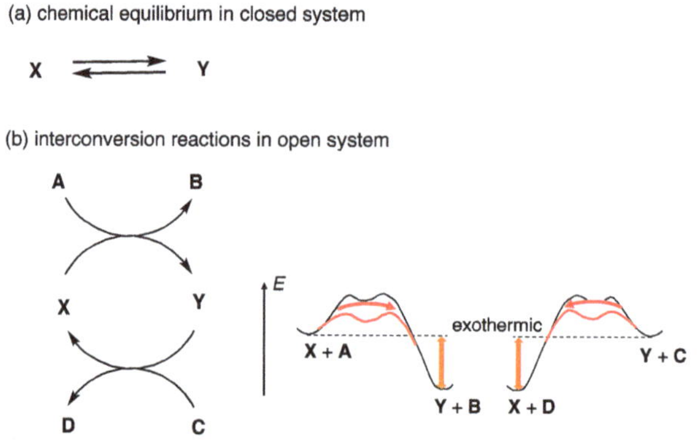

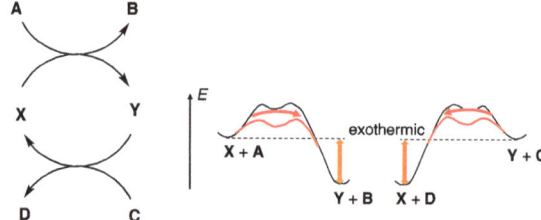

Figure 4. (**a**) Chemical equilibrium in a closed system between **X** and **Y**. (**b**) Interconversion reaction in an open system between **X** and **Y**, in which the relative thermodynamic stability of substrates and products is inverted and induces **X** + **A** → **Y** + **B** and **Y** + **C** → **X** + **D** reactions.

2. Rhodium-Catalyzed Exchange Reactions of Disulfides

Disulfides can exchange organothio groups under acidic or basic conditions and when exposed to heat or photoirradiation. We found that disulfide exchange reactions proceed efficiently in the presence of a rhodium complex [32]. A mixture of RhH(PPh$_3$)$_4$, sulfonic acid, and phosphine promotes catalysis of the disulfide exchange reaction, which proceeds rapidly under acetone reflux (Scheme 1). Dibutyl disulfide and bis(2-benzoyloxyethyl) disulfide were reacted in refluxing acetone in the presence of RhH(PPh$_3$)$_4$ (3 mol%), trifluoromethanesulfonic acid (6 mol%), and (p-tol)$_3$P (12 mol%), and 2-benzoyloxyethyl butyl disulfide was obtained in 49% yield. The reaction is applicable to both aliphatic and aromatic disulfides, and it proceeds at a much lower temperature than in the conventional heating method. A chemical equilibrium was rapidly reached within 15 min, which included two symmetric disulfides and one asymmetric disulfide at a statistical 1:1:2 ratio. Chemical equilibrium was confirmed by the reverse reaction of an unsymmetric disulfide. The method is applicable to the exchange of disulfides, diselenides/ditellurides, and other heteroatom compounds of group 16 elements. The proposed mechanism involves the initial oxidative addition of a low-valent rhodium complex and a disulfide. The subsequent oxidative addition of another disulfide or ligand exchange of the organothio group occurs, and the disulfide is liberated by reductive elimination with the regeneration of the catalyst.

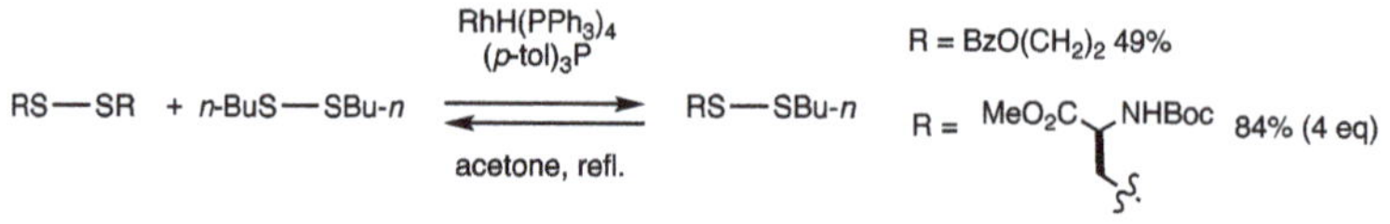

Scheme 1. Disulfide exchange reaction.

When one product is removed from the solution of the disulfide exchange reaction, the chemical equilibrium can be shifted (Figure 5a) [33]. We developed a silica nanoparticle precipitation method, in which nanoparticles precipitated from solution with concomitant molecular recognition and adsorption of molecules. Silica (*P*)-nanoparticles of 70 nm mean diameter grafted with (*P*)-helicene were employed. When butyl (*R*)-(hydroxyphenylmethyl) disulfide (*R*)-**45** was treated with RhH(PPh$_3$)$_4$ (20 mol%), trifluoromethanesulfonic acid (40 mol%), and (p-tol)$_3$P (80 mol%) in chlorobenzene for 24 h in the presence of silica (*P*)-nanoparticles, precipitates containing (*R*,*R*)-bis(hydroxyphenylmethyl) disulfide (*R*, *R*)-**43** (27%) and (*R*)-**45** (3%) were formed (Figure 5b). The high preference for (*R*, *R*)-**43** is notable, and no dibutyl disulfide **44** was contained in the precipitates. The solution phase contained (*R*, *R*)-**43** (6%), (*R*)-**45** (30%), and **44** (33%). The composition of disulfides in the precipitates deviates considerably from that of the initial chemical equilibrium with (*R*, *R*)-**43**:(*R*)-**45**:**44** = 1:2:1. It should also be noted that most of the butylthio group remained in solution.

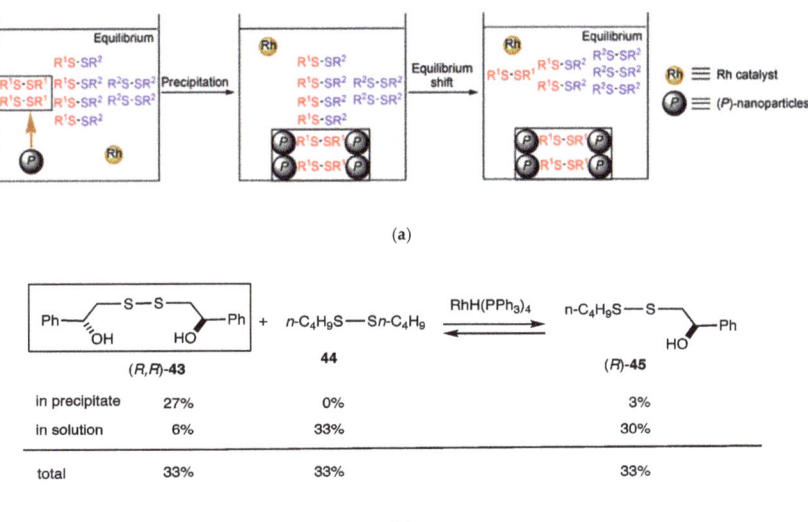

(a)

(b)

Figure 5. (**a**) Equilibrium shift in disulfide exchange reaction induced by the precipitation of silica (*P*)-nanoparticles. (**b**) Equilibrium shift in the reaction of (*R*, *R*)-**43**, **44**, and (*R*)-**45**.

The disulfide exchange reaction of hydrophilic disulfides occurs in water when using the $RhCl_3$ catalyst (Scheme 2) [34]. When glutathione disulfide and glycolic acid disulfide (4 equivalents) were treated with $RhCl_3$ (10 mol%) in water at 40 °C for 1 h, methylthiolated glutathione was obtained in 81% yield. The reverse reaction confirmed the involvement of chemical equilibrium. In addition, the composition under chemical equilibrium could be changed by the addition of a disulfide, which indicated that catalysis occurred after reaching chemical equilibrium. This method can be applied to the reaction of dimethyl disulfide, which is not water-soluble. A two-phase system of glutathione disulfide in water and excess dimethyl disulfide was treated with $RhCl_3$ (10 mol%) at 40 °C for 1 h, and methylthiolated glutathione was obtained in 40% yield.

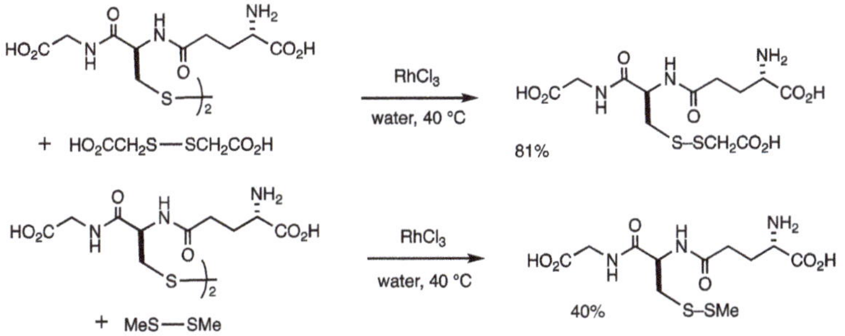

Scheme 2. Disulfide exchange reaction of glutathione in water.

The RhCl$_3$-catalyzed method was applied to insulin containing three S-S bonds, and the reaction with excess thioglycolic acid preferentially exchanged the disulfide at the 7 positions in both A and B chains (Scheme 3) [35]. The product was obtained in 30% yield at room temperature for 1 h using RhCl$_3$ (300 mol%) with recovery of insulin in 70% yield. The rhodium-catalyzed method in water tolerates various functional groups, including amides, amines, carboxylic acids, alcohols, and phenols, because of the higher affinity of rhodium for sulfur atoms in the presence of nitrogen and oxygen functional groups.

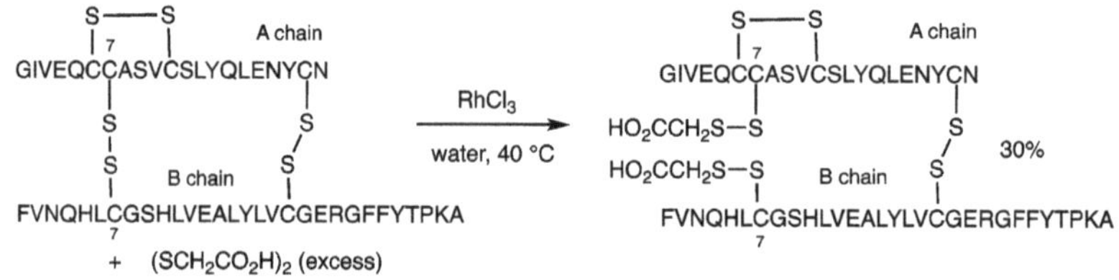

Scheme 3. Disulfide exchange reaction of insulin.

3. Rhodium-Catalyzed Substitution Reactions Using Disulfides

The S-S bonds of disulfides are reversibly cleaved under rhodium catalysis, as indicated by the disulfide exchange reaction. Rhodium complexes also cleave C-S and C-H bonds in organic compounds, and combinations of these bond cleavage reactions provide various chemical transformations for the synthesis of organosulfur compounds, including thioesters, α-thioketones, 1-thioalkynes, aryl sulfides, and dithiophosphinates.

3.1. Substitution Reactions of Thioesters

Thioesters are excellent substrates for rhodium-catalyzed reactions and provide various acyl derivatives by their reactions with different substrates. Organothio exchange reaction of thioesters occurs with disulfides (Scheme 4) [36]. S-Octyl benzothioate and bis(2-ethoxyethyl) disulfide (4 equivalents) were reacted in the presence of RhCl(PPh$_3$)$_3$ (2.5 mol%) at 3-pentanone reflux for 1.5 h, and S-(2-ethoxyethyl) benzothioate was obtained in 87% yield. Chemical equilibrium was determined by the reverse reaction and formation of a statistical 1:2:1 mixture employing 1 equivalent of disulfides. It was also observed that the reaction of S-methyl thioesters and disulfides (4 equivalents) gave higher yields of the exchange products under 1,2-dichlorobenzene reflux because of the removal of volatile dimethyl disulfide from the reaction mixture. These results indicate facile cleavage of C-S bonds in thioesters by rhodium catalysis.

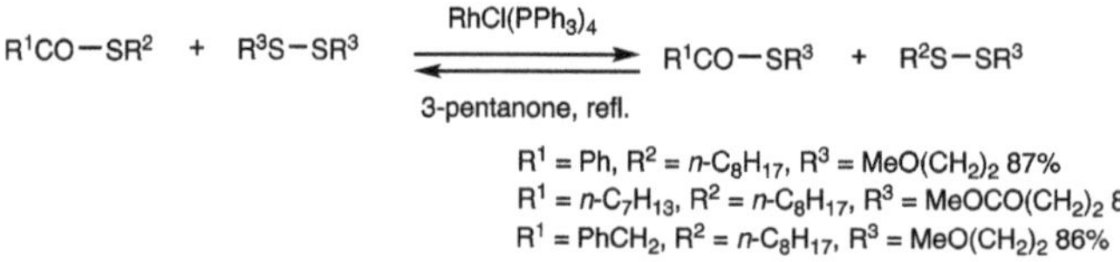

Scheme 4. Organothio exchange reaction of thioesters.

Acid fluorides show high reactivity under rhodium-catalyzed conditions (Scheme 5) [37]. When benzoyl fluoride, bis(p-tolyl) disulfide, and triphenylphosphine were reacted in the presence of RhH(PPh$_3$)$_4$ (1 mol%) and 1,2-diphenylphosphinoethane (dppe) (2 mol%) in refluxing THF for 2 h, S-(p-tolyl) benzothioate was obtained in 100% yield. Formally, the reaction provides unstable sulfenyl fluoride with disulfide regeneration. The role of triphenylphosphine is to trap fluorides to form phosphine difluoride with disulfide regeneration, which results in an exothermic reaction suitable for catalysis.

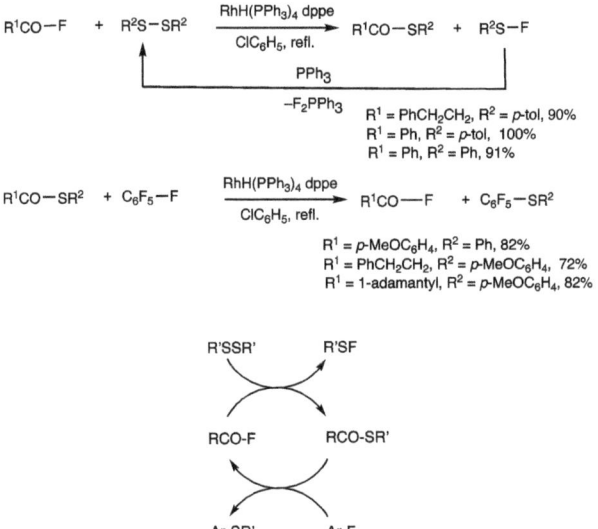

Scheme 5. Interconversion between acid fluorides and thioesters.

The rhodium-catalyzed method was also employed in the synthesis of acid fluorides from thioesters (Scheme 5) [37]. When S-(p-tolyl) benzothioate and hexafluorobenzene were reacted in the presence of RhH(PPh$_3$)$_4$ (2.5 mol%) and dppe (5 mol%) in refluxing chlorobenzene, benzoyl fluoride was obtained in 94% yield in addition to 1,4-di(p-tolylthio)-2,3,5,6-tetrafluorobenzene. This is a noteworthy fluorinating reaction using stable neutral aromatic fluorides. Chemical equilibrium also occurs between an acid fluoride and a thioester in a closed system under rhodium catalysis.

The above results indicate involvement of interconversion reactions in an open system between thioester and acid fluoride catalyzed by rhodium: Acid fluorides are converted to thioesters by adding disulfides, and thioesters are converted to acid fluorides by adding hexafluorobenzene. The addition of external reagents inverts the relative thermodynamic stability and promotes the interconversion reactions (Figure 4). It was considered that a R'SSR' + ArF → Ar-SR' + R'S-F reaction may also occur; this reaction is described later (Scheme 12).

Rhodium-catalyzed reactions of disulfides can be applied to diselenides (Scheme 6) [38]. Bis(2-pyridyl) diselenide and 1-adamantanecarbonyl fluoride were reacted in the presence of RhH(PPh$_3$)$_4$ (2.5 mol%) and dppe (5 mol%) in refluxing chlorobenzene, and 1-adamantanecarbonyl 2-pyridylselenoester was obtained in 88% yield. This method was used for the synthesis of heteroaryl selenoesters, which are generally less stable than aryl thioesters. The heteroaryl compounds can be used for the synthesis of novel organoselenium compounds.

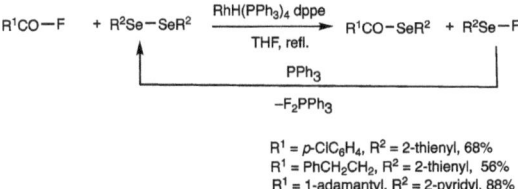

Scheme 6. Synthesis of selenoesters.

3.2. Substitution Reactions of 1-Alkynes

Rhodium complexes can activate C-H bonds of organic molecules, and this reaction is employed here for thiolation reactions using disulfides [30]. The reaction of a S-S bond in a disulfide and a C-H bond in an organic compound formally provides an organosulfur compound with a C-S bond and a thiol with a S-H bond. The thiol can be oxidized to a disulfide in the presence of oxygen, a reaction that is also catalyzed by rhodium complexes.

When 1-(triethylsilyl)-acetylene was treated with dibutyl disulfide in the presence of $RhH(PPh_3)_4$ (2 mol%) and diphenylphosphinoferrocene (dppf) (3 mol%) for 1 h in refluxing acetone, 1-butylthio-2-triethylsilylacetylene was obtained in 80% yield, which was accompanied by a thiol (Scheme 7) [39]. The reaction of a thiol and a 1-thioacetylene formed the original 1-alkyne under argon atmosphere. 1-Alkynes/disulfides and 1-thioalkyne/thiols are under chemical equilibrium in the presence of rhodium catalysis. The rhodium complex also rapidly oxidizes thiols to disulfides and water in the presence of a trace amount of oxygen (Scheme 7 and Scheme 23). In combination with the oxidation reaction under air, the thiolation reaction of 1-alkynes proceeds in an energetically downhill manner to provide 1-thioalkynes in higher yields. The interconversion reactions proceed between 1-alkynes and 1-thioalkynes.

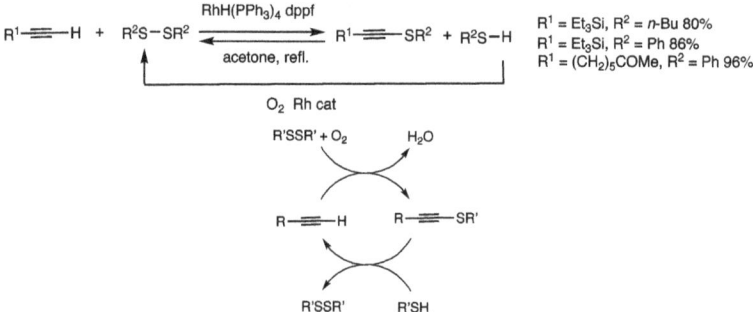

Scheme 7. Organothiolation reaction of 1-alkynes.

A rhodium complex catalyzes the organothio exchange reaction of 1-thioalkynes with disulfides, the observation of which confirmed the C-S bond cleavage by rhodium catalysis (Scheme 8) [39]. In principle, diverse 1-thialkynes can be obtained by this method using different disulfides.

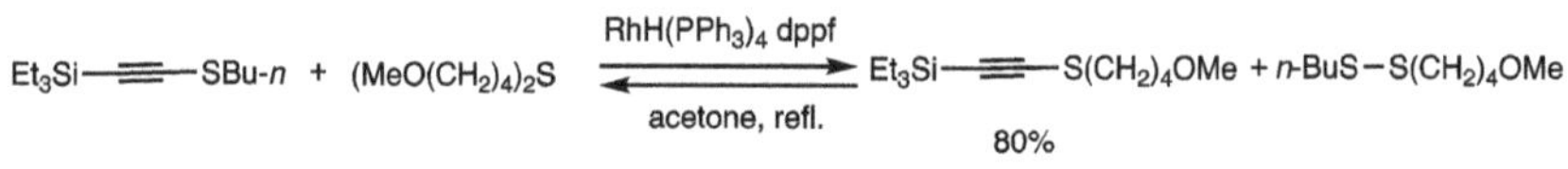

$$Et_3Si-\!\!\equiv\!\!-SBu\text{-}n + (MeO(CH_2)_4)_2S \xrightarrow[\text{acetone, refl.}]{RhH(PPh_3)_4\ dppf} Et_3Si-\!\!\equiv\!\!-S(CH_2)_4OMe + n\text{-}BuS-S(CH_2)_4OMe$$
$$80\%$$

Scheme 8. Organothio exchange reaction of 1-thioalkyne.

3.3. Substitution Reactions of Active Methylene Compounds

Active methylene compounds, which include nitroalkanes, malonate, and benzyl ketones, with acidic hydrogen atoms are thiolated with disulfides under rhodium catalysis (Scheme 9) [40]. When 1-nitropentane was treated with bis(*p*-chlorophenyl) disulfide in *N*, *N*-dimethylacetamide (DMA) in air at room temperature for 3 h in the presence of $RhH(PPh_3)_4$ (5 mol%) and dppe (10 mol%), 2-(*p*-chlorophenyl) nitropentane was produced in 50% yield. The reaction without air is under chemical equilibrium, and α-thiolated nitroalkanes are thermodynamically unfavorable. Accordingly, treatment of an α-thiolated nitroalkane with a thiol quantitively provided the original nitroalkane. Air/oxygen converts thiols to disulfides and water in an exothermic reaction. Diethyl malonate and 1,2-diphenyl-1-ethanone were also reacted with aromatic disulfides in air to provide organothiolated compounds.

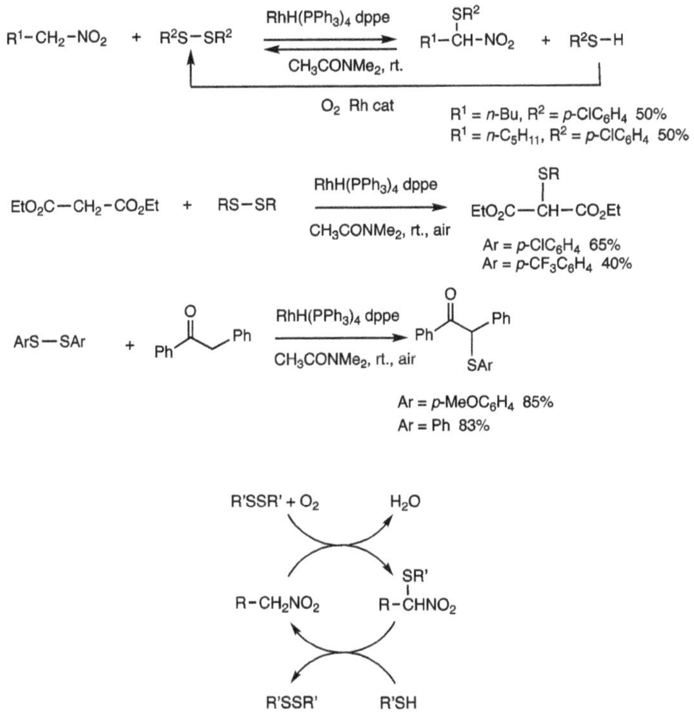

Scheme 9. Organothiolation reactions of active methylene compounds.

3.4. Substitution Reactions of Ketones and Heteroarenes

α-Thioketones efficiently undergo C-S substitution reactions in the presence of rhodium catalysts. In combination with the C-H cleavage reaction catalyzed by rhodium complexes, organothiolation of various organic compounds, including ketones and heteroarenes, occurs.

It was determined that organothio exchange reactions of α-thioketones with disulfides proceeded with the involvement of C-S bond cleavage by rhodium catalysis (Scheme 10) [41]. When an α-phenylthioacetophenone was treated with bis(3-methoxypropyl) disulfide (3 equivalents) in the presence of RhH(PPh$_3$)$_4$ (1 mol%) and dppe (2 mol%) in refluxing THF for 1–2 h, α-(3-methoxypropyl) acetophenone was obtained in 82% yield. The reverse reaction under the same conditions indicated the involvement of chemical equilibrium. The exchange reactions of organothio groups using different disulfides under rhodium catalysis provide diverse derivatives starting from a single α-thioketone.

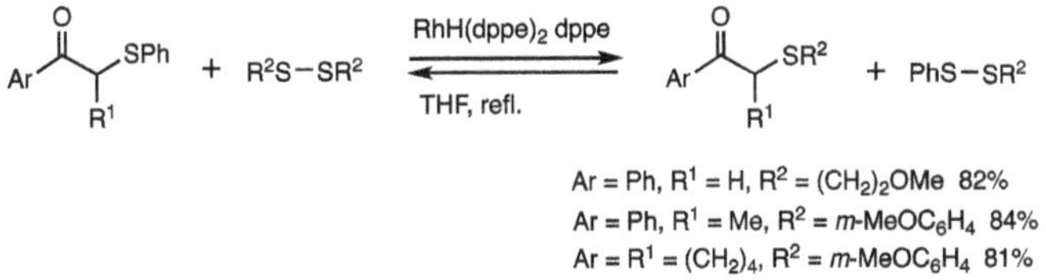

Scheme 10. Organothio exchange reaction of α-thioketones.

α-Thioketones can be used as organothiolating reagents of organic compounds via organorhodium intermediates with C-Rh-S subunits (Scheme 11). A methylthio transfer reaction occurs between different ketones at the α-position, in which a catalytic amount of dimethyl disulfide significantly promotes the reaction [42]. When α-methylthio-*p*-cyanoacetophenone and 1,2-diphenylethanone were reacted in refluxing THF for 3 h in the presence of RhH(PPh$_3$)$_4$ (4 mol%), dppe (8 mol%), and dimethyl disulfide (12 mol%), 2-methylthio-1,2-diphenylethanone was obtained in 68% yield. The methylthio group moved from α-methylthio-*p*-cyanoacetophenone to 1,2-diphenylethanone under chemical equilibrium. This method is applied to cyclic α-phenyl ketones with acidic α-protons. The reaction of 2-phenylthio-4-(*t*-butyl) cyclohexanone and α-methylthio-*p*-chloroacetophenone gave a product with an axial methylthio group, and the cleavage of the phenylthio group was slow under these conditions [43].

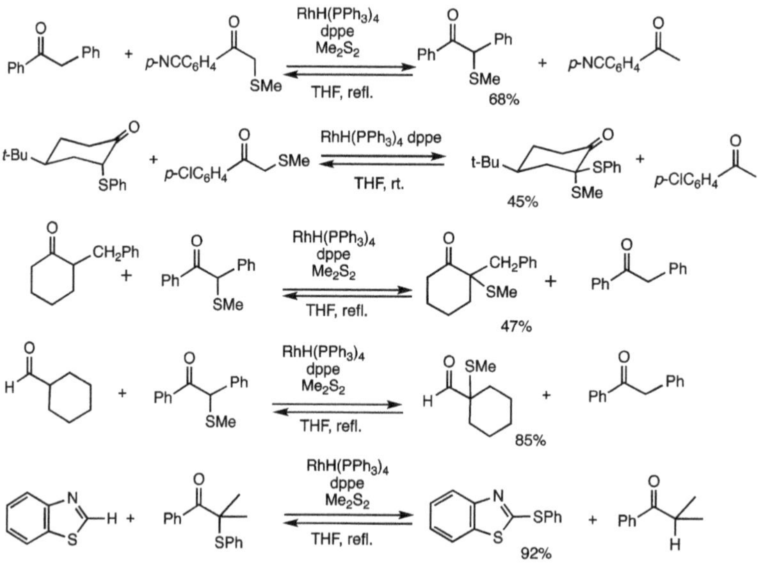

Scheme 11. Organothiolation reactions of ketones and benzothiazole.

2-Methylthio-1,2-diphenylethanone is effective in the α-methylthiolation reactions of ketones with less acidic α-protons compared with α-methylthio-*p*-cyanoacetophenone [44]. Reaction of 2-benzylcyclohexanone and 2-methylthio-1,2-diphenylethanone in the presence of RhH(PPh$_3$)$_4$ (4 mol%), dppe (8 mol%), and dimethyl disulfide (12 mol%) in refluxing THF for 3 h gave 2-benzyl-2-methylthiocyclohexanone in 47% yield. The regioselectivity indicated the important role of the enol form of 2-methylcyclohexanone. The method is also applied to α-methylthiolation of aldehydes. In these α-thiolation reactions, disulfides alone did not give satisfactory results, and α-thioketones were employed as the donors. A favorable chemical equilibrium may be involved in forming ketones from α-thioketones rather than forming thiols from disulfides. Possible roles of dimethyl disulfide are the generation of reactive rhodium species with sulfur ligands and/or the transient formation of α-methylthio ketones.

Phenylthiolation reactions of aromatic hydrogen atoms in benzo-fused heteroaromatic compounds were conducted using α-phenylthioisobutyrophenone, which provided higher yields of the products when compared with 2-methylthio-1,2-diphenylethanone [45]. This result may be due to the thermodynamically favorable nature of chemical equilibrium to form a ketone with a less acidic α-proton. 1,3-Benzothiazole and α-methylthioisobutyrophenone were reacted in chlorobenzene reflux for 3 h in the presence of RhH(PPh$_3$)$_4$ (4 mol%) and dppe (8 mol%), which provided 2-phenylthio-1,3-benzothiazole in 92% yield. The methylthiolation reaction of benzothiazole provided α-methythioisobutyrophenone in the presence of dimethyl disulfide, albeit in lower yields.

These results indicate that methylthio groups can hop between α-protons of ketones under rhodium catalysis (Figure 6). This is analogous to the α-proton exchange reactions between ketones under acidic or basic conditions.

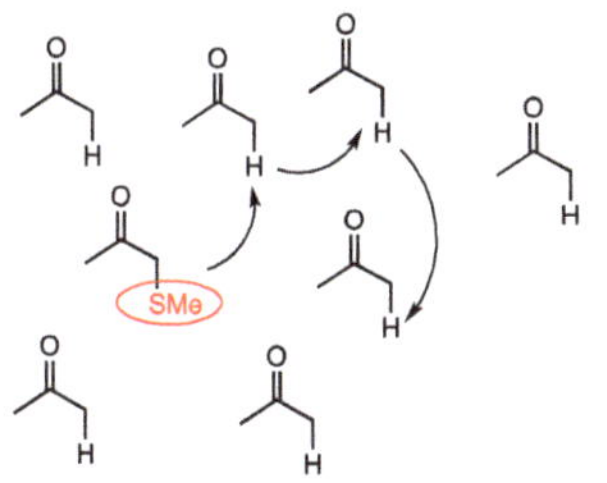

Figure 6. Rhodium-catalyzed hopping of methylthio groups among ketones at α-positions.

3.5. Substitution Reactions of Aromatic Fluorides

Organofluorine compounds are highly reactive under rhodium catalysis, as noted in the reaction of acyl fluorides to form thioesters (Scheme 5). This behavior may be due to the high reactivity of rhodium fluoride intermediates, which is consistent with the notably high reactivity of RhF(PPh$_3$)$_3$ [46]. Accordingly, C-F bonds in fluorobenzenes were converted to C-S bonds via reaction with disulfides (Scheme 12). When 1-bromo-4-chloro-3-fluorobenzene was reacted with bis(p-tolyl) disulfide and triphenylphosphine in the presence of RhH(PPh$_3$)$_4$ (0.25 mol%) and 1,2-(diphenylphosphino) benzene (dppBz) (0.5 mol%) under chlorobenzene reflux for 3 h, 5-bromo-2-chlorophenyl p-tolyl sulfide was obtained in 72% yield [47]. The reaction proceeded selectively at the fluoride atom without affecting chloride and bromide atoms. Triphenylphosphine was employed to trap fluorides by forming phosphine difluoride, which resulted in an exothermic reaction. The reaction of perfluorobenzenes showed notable selectivity in the substitution reaction, and thiolation occurred at the p-positions, which was named the p-difluoride rule. The reaction of hexafluorobenzene and bis(p-tolyl) disulfide gave 1,4-bis(p-tolyl)-2,3,5,6-tetrafluorobenzene, in which no monothiolated product was detected.

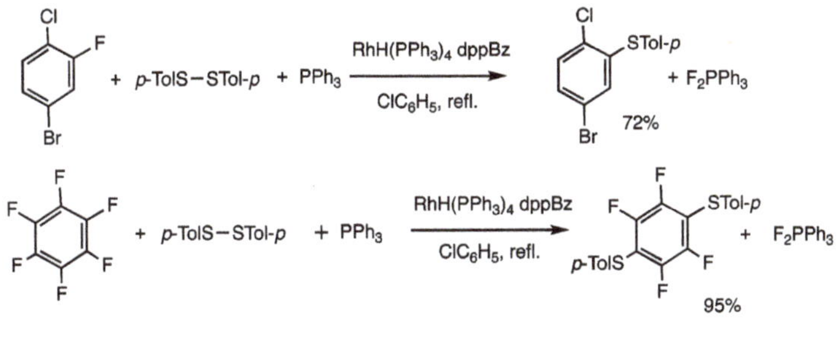

Scheme 12. Substitution reaction of aromatic fluorides.

3.6. Substitution Reactions of Organophosphorus Compounds

Organophosphorus compounds efficiently react with disulfides in rhodium-catalyzed substitution reactions, owing to the strong bonding of sulfur and phosphorus atoms. Cleavage of the P-S bond was determined by organothio exchange reactions of dithiophosphinates and disulfides (Scheme 13) [48]. When phenyl dimethyldithiophosphinate and bis(4-methoxybutyl) disulfide were reacted in the presence of RhH(PPh$_3$)$_4$ (2 mol%) and dppe (4 mol%) under acetone reflux for 0.5 h, the corresponding dithiophosphinate was obtained in 74% yield.

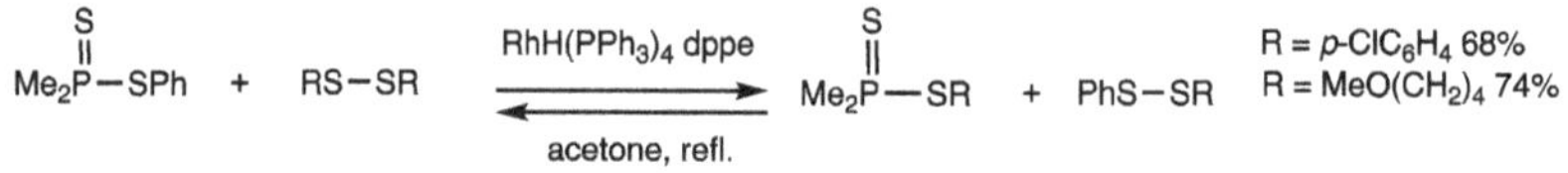

Scheme 13. Organothio exchange reaction of dithiophosphinates.

Reactions of acylphosphine sulfides and disulfides provided dithiophosphinates and thioesters, which involved the formation of C-S and P-S bonds from C-P and S-S bonds (Scheme 14) [49]. When diethyl(p-dimethylaminobenzoyl) phosphine sulfide was reacted with di(undecyl) disulfide in the presence of RhH(PPh$_3$)$_4$ (2 mol%) and dppe (4 mol%) in refluxing THF for 3 h, the corresponding thioester was obtained in 90% yield, along with dithiophosphonate.

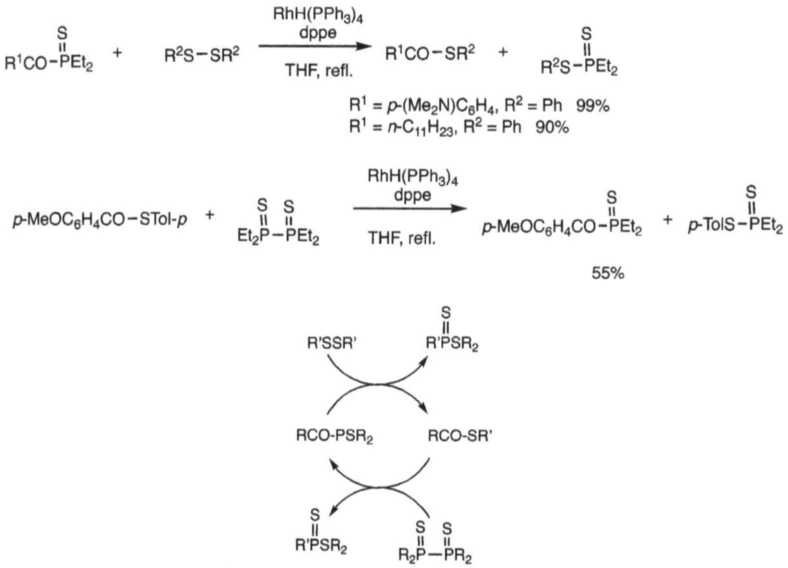

Scheme 14. Interconversion reactions between acylphosphine sulfides and thioesters.

Thioesters were converted to acylphosphine sulfides by reacting with diphosphine sulfides (Scheme 14) [49]. When S-(p-tolyl) p-methoxybenzothioate and tetraethyldiphosphine disulfides were reacted in refluxing chlorobenzene for 6 h in the presence of RhH(PPh$_3$)$_4$ (5 mol%) and 1,2-(diethylphosphino) ethane (depe) (10 mol%), acylphosphine sulfide was obtained in 55% yield. Thus, interconversion reactions in an open system occur between thioesters and acylphosphines in the presence of appropriate reagents, which are catalyzed by rhodium complexes; i.e., organosulfur compounds and organophosphorus compounds may be interconverted.

On the basis of analysis of interconversion reactions in open systems, it was suggested that the treatment of diphosphine disulfides and disulfides provides dithiophosphinates via the exchange of P-P bonds and S-S bonds (Figure 4) [48]. When dioctyl disulfide and tetramethyldiphosphine disulfide were reacted under acetone reflux for 0.5 h in the presence of RhH(PPh$_3$)$_4$ (1.5 mol%) and dppe (3 mol%), thiophosphinate was obtained in 97% yield (Scheme 15). The reaction is irreversible, owing to the formation of strong P-S bonds from weak P-P bonds. The reaction is also applicable to diselenides.

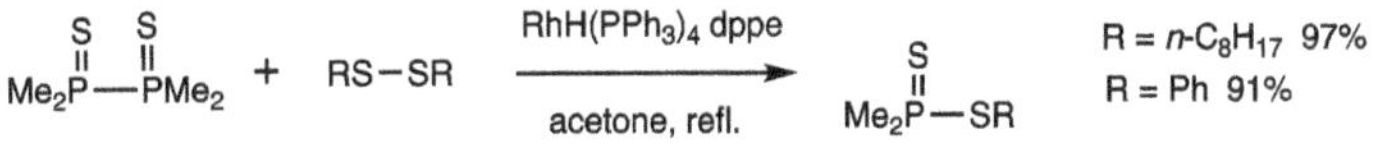

Scheme 15. Exchange reaction of diphosphine disulfides and disulfides.

The reaction of hydrophilic disulfides and diphosphonates, which are also water-soluble, proceeds in water in the presence of $RhCl_3$ (Scheme 16) [50]. Glutathione disulfide was reacted with tetramethyl diphosphonate in water at 40 °C for 36 h in the presence of $RhCl_3$ (10 mol%), and glutathione phosphonate was obtained in 77% yield.

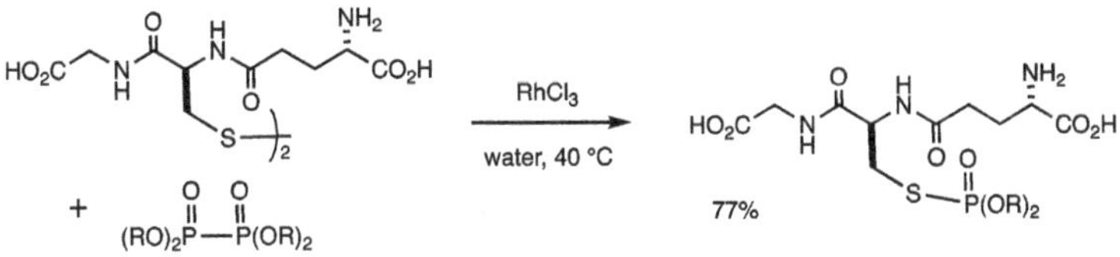

Scheme 16. Exchange reaction of diphosphine disulfides and disulfides in water.

Polyphosphines are compounds containing phosphorus atoms with two P-P bonds and one organic group, and they exhibit reactivities different from those of diphosphines, which contain phosphorus atoms with one P-P bond and two organic groups. Polyphosphines undergo substitution reactions with disulfides, which involve the exchange of P-P and S-S bonds (Scheme 17) [51]. Pentaphenylcyclopentaphosphine was reacted with dihexyl disulfide in the presence of $RhH(dppe)_2$ (5 mol%) in THF reflux for 15 min, which produced dihexyl phenylphosphonodithionite in 94% yield. When cyclic disulfides are employed, novel heterocyclic phosphonodithinites are obtained. The intermediate formation of a diphosphene–rhodium complex was shown by spectroscopic and MS analyses.

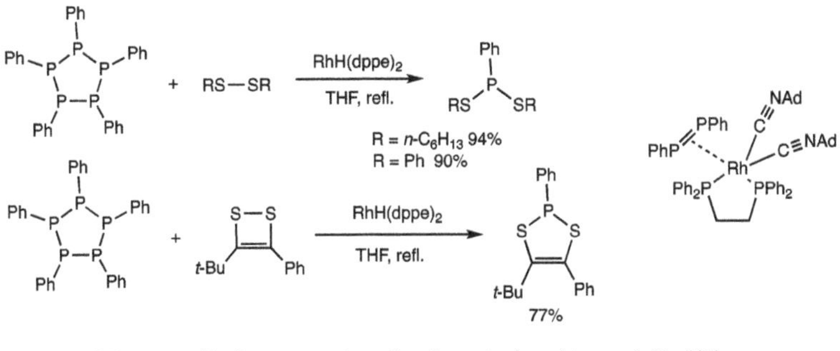

Scheme 17. Exchange reaction of cyclopentaphosphine and disulfides.

A rhodium catalyst promoted the cleavage of C-P bonds in heteroarylphosphine sulfides and exchange with disulfides (Scheme 18) [52]. When (2-benzothiazolyl) dimethylphosphine sulfide was reacted with dioctyl disulfide in the presence of $RhH(PPh_3)_4$ (10 mol%) and 1,2-bis (dimethylphosphino)benzene (dmppBz) (20 mol%) in refluxing chlorobenzene for 3 h, 2-(octylthio) benzothiazole was obtained in 46% yield.

The reverse reaction converts heteroaryl sulfides to heteroarylphosphine sulfides in the presence of diphosphine disulfides (Scheme 18) [52,53]. When 2-(p-tolylthio)-1,3-benzothiazole and tetraethyldiphosphine disulfides were reacted in refluxing chlorobenzene for 6 h in the presence of $RhH(PPh_3)_4$ (5 mol%) and 1,2-(diethylphosphino) ethane (depe) (10 mol%), 2-(diethylphosphino)-

1,3-benzothiazole sulfide was obtained in 49% yield. This is another example of interconversion reactions in an open system, in which organophosphorus and organosulfur compounds are interconverted.

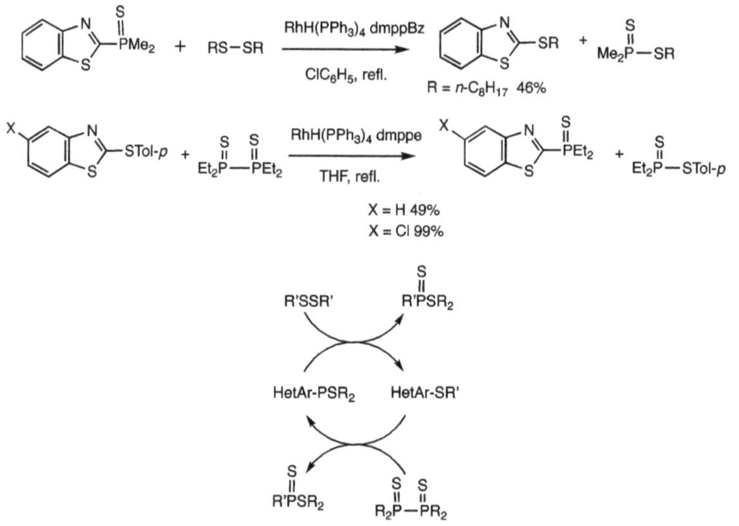

Scheme 18. Interconversion reactions of heterocyclic phosphine sulfides and heterocyclic sulfides.

4. Rhodium-Catalyzed Insertion Reactions Using Disulfides

Insertion reactions of disulfides and alkynes/alkenes are generally exothermic, involving the conversion of high-energy unsaturated compounds to low-energy less unsaturated compounds. Previously reported palladium-catalyzed insertion reactions of disulfides with alkynes in general employed diaryl disulfides [21,23]. Rhodium-catalyzed insertion of 1-akynes may be applied to aliphatic disulfides, which are less reactive than aromatic disulfides (Scheme 19) [54].

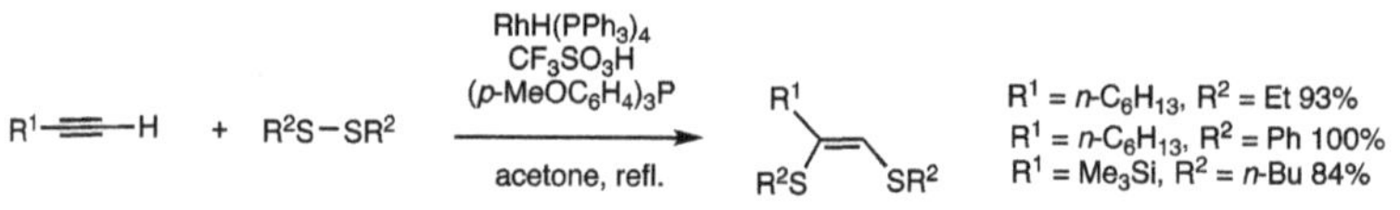

Scheme 19. Insertion reaction of disulfides and 1-alkynes.

When 1-octyne and diethyl disulfide were reacted in refluxing acetone for 10 h in the presence of RhH(PPh$_3$)$_4$ (4 mol%), trifluoromethanesulfonic acid (4 mol%), and (p-MeOC$_6$H$_4$)$_3$P (14 mol%), 1,2-bis(alkylthio)-1-alkene with the Z-configuration was obtained in 93% yield. The proposed mechanism involves the oxidative addition of low-valent rhodium and a disulfide, followed by the transfer of two organothio groups to 1-alkyne.

A mixture of a disulfide and a diselenide predominantly provides 1-seleno-2-thioalkene, which is accompanied by minor amounts of other insertion products (Scheme 20) [55]. When 1-octyne, diphenyl disulfide, and diphenyl diselenide were reacted in refluxing acetone for 4 h in the presence of RhH(PPh$_3$)$_4$ (5 mol%) and 1,4-(diphenylphosphino) butane (dppb) (10 mol%), 2-butylthio-1-butylseleno-1-octene was obtained in 72% yield. The reaction is also applicable to dialkyl disulfides/diselenides. The reaction is under chemical equilibrium of disulfide/diselenide exchange, in which selenosulfides are selectively transferred to 1-alkynes.

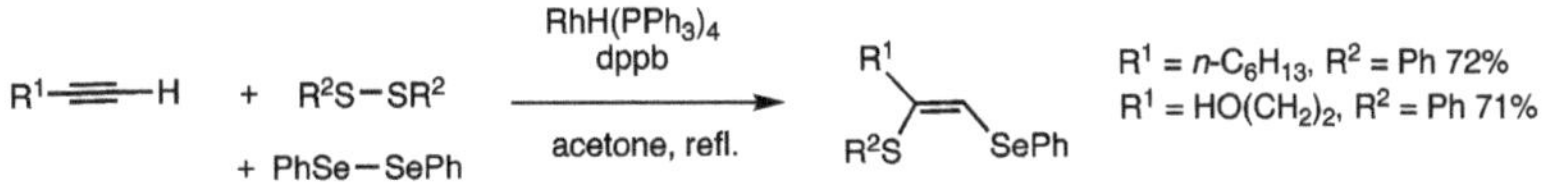

Scheme 20. Insertion reaction of disulfides/diselenides and 1-alkynes.

The reaction of allene in place of 1-alkyne provides a different variety of products (Scheme 21) [56]. When 1,2-octadiene and dibutyl disulfide were reacted in refluxing acetone for 2 h in the presence of RhH(PPh$_3$)$_4$ (3 mol%), trifluoromethanesulfonic acid (3 mol%), and (p-tol)$_3$P (12 mol%), 2-(butylthio)-1,3-octadiene and (E)-2-(butylthio)-2-octene were both obtained in 47% yields. Two organothio groups are transferred to two allene molecules with concomitant hydride transfer. This reaction is also applicable to diselenides.

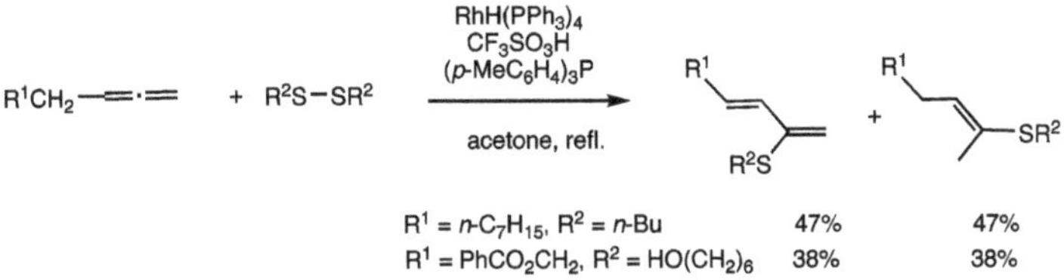

Scheme 21. Insertion reaction of allenes and disulfides.

The α-insertion reaction of carbon monoxide with disulfide occurred to provide dithiocarbonate (Scheme 22) [36]. When dibutyl disulfide was treated with carbon monoxide at 30 atm in toluene at 180 °C for 24 h in the presence of RhH(PPh$_3$)$_4$ (10 mol%) and dppe (20 mol%), dibutyl S, S′-dithiocarbonate was obtained in 15% yield. This reaction is under equilibrium and favors disulfide and carbon monoxide at ambient pressure as well as at 30 atm. Rhodium-catalyzed α-insertion reactions of carbenoids and disulfides have been reported [57,58].

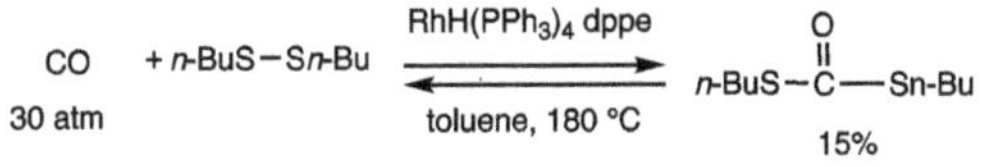

Scheme 22. Insertion reaction of carbon monoxide and disulfide.

5. Rhodium-Catalyzed Reduction/Oxidation Reactions of Disulfides

Interconversion of a disulfide and two thiols is an important chemical reaction in organosulfur chemistry. Various methods have been reported for the reduction of disulfides to thiols using different reducing reagents. Hydrogenation may be the most convenient in terms of availability of reducing reagent and simple operations, which is also achieved by rhodium catalysis (Scheme 23). When di(octyl) disulfide was treated with hydrogen at 1 atm in the presence of RhH(PPh$_3$)$_4$ (0.5 mol%) in refluxing toluene for 0.5 h, 1-octanethiol was obtained in 90% yield [59]. Care to deactivate a rhodium complex is critical during quenching of the reaction because thiols are rapidly converted to disulfide under air in the presence of the rhodium complex. Metal-catalyzed hydrogenolysis of disulfides to thiols has been rare, which may be because of catalyst poisoning by thiols.

The reverse reaction, oxidation of thiols to disulfides, is also catalyzed by the rhodium complex using oxygen as the oxidation reagent (Scheme 23) [59]. Treatment with 1-octanethiol under oxygen at 1 atm in methanol at 0 °C for 1 h in the presence of RhH(PPh$_3$)$_4$ (0.1 mol%) and 1,4-bis(diphenylphosphino)butane (dppb) (0.2 mol%) provided dioctyl disulfide in 93% yield.

These interconversion reactions in open systems can be a convenient method for the development of molecular switching functions, which employ hydrogen as the reducing reagent and oxygen as the oxidizing reagent.

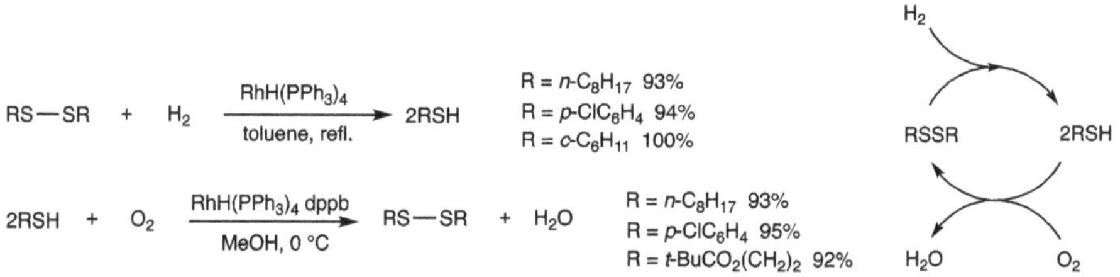

Scheme 23. Interconversion reactions of disulfides and thiols.

6. Rhodium-Catalyzed Reactions of Sulfur

Sulfur is a readily available source of organosulfur compounds in organic synthesis. Conventional methods using sulfur generally employ the formation of thiyl radicals, which are generated by heating above the melting point of sulfur (115 °C) [30]. Several previously reported metal-catalyzed methods employ high temperatures, which likely involve radical- and metal-catalyzed mechanisms and accordingly are not easy to analyze and control. To secure the effect of metal catalysis, we conducted rhodium-catalyzed reactions of sulfur well below the melting point. The reactivity of sulfur can be different from disulfide, and Rh-S-S intermediates formed from sulfur can exhibit different reactivities from Rh-S-C intermediates formed from disulfides because of the presence of the adjacent weak S-S bonds in the latter.

Introduction of sulfur atoms between S-S bonds in disulfides provides polysulfides, and such reactions occur above the melting point and involve radical mechanisms. Rhodium-catalyzed reactions proceed at much lower temperatures [60]. In the presence of RhH(PPh$_3$)$_4$ (2.5 mol%) and 1,2-bis(diphenylphosphino)ethene (dppv) (5 mol%), dibutyl trisulfide and sulfur were reacted in acetone at room temperature for 5 min, and dibutyl tetrasulfide, pentasulfide, and hexasulfide were obtained in 39, 14, and 8% yields, respectively, accompanied by higher homologs (Scheme 24). The reaction occurred rapidly and provided a mixture of polysulfides. Aliphatic disulfides were not reactive under these conditions. In contrast, diaryl disulfides effectively reacted with sulfur to provide polysulfides, including trisulfides, tetrasulfides, and pentasulfides.

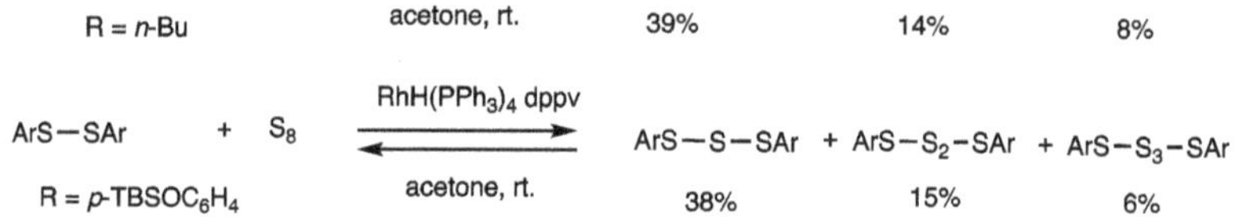

Scheme 24. Insertion reaction of sulfur and disulfides.

Thioisonitriles were synthesized by sulfuration of isonitriles under rhodium-catalyzed conditions, which resulted in the 1,1-insertion of sulfur at the nitrogen atom of the C=N bond (Scheme 25) [61]. The reaction of cyclohexylisonitrile and sulfur under acetone reflux for 2 h in the presence of RhH(PPh$_3$)$_4$ (1 mol%) provided thioisonitrile in 91% yield in 2 h. The reaction is faster than the known molybdenum method developed by Bargon [62]. Trisulfides and tetrasulfides can also react under rhodium catalysis. An induction period was observed with an initially slow reaction for 40 min followed by a rapid reaction completed in 100 min. The phenomenon was ascribed to the slow formation of active sulfur species because preheating sulfur in acetone for 1.5 h eliminated the induction period.

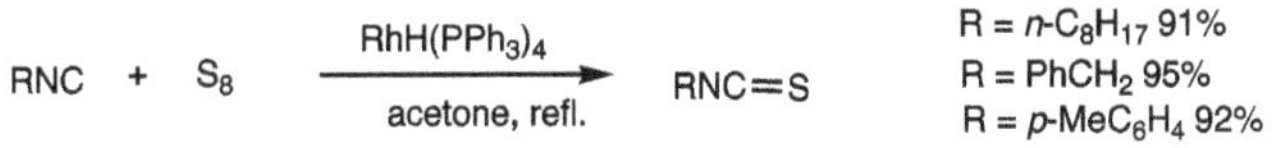

$$RNC + S_8 \xrightarrow[\text{acetone, refl.}]{\text{RhH(PPh}_3)_4} RNC=S \quad \begin{array}{l} R = n\text{-}C_8H_{17}\ 91\% \\ R = PhCH_2\ 95\% \\ R = p\text{-}MeC_6H_4\ 92\% \end{array}$$

Scheme 25. Insertion reaction of sulfur and thioisonitriles.

Symmetric diaryl sulfides are synthesized from perfluorobenzene and sulfur involving C-F bond activation, which showed reactivity and the *p*-difluoride rule similar to disulfides (Scheme 26) [63]. Treatment of 4-phenylthiopentafluorobenzene, sulfur, and tributylsilane in DMF at room temperature in the presence of RhH(PPh$_3$)$_4$ (5 mol%) and dppBz (10 mol%) provided bis(4-cyano-2,3-5,6-tetrafluorophenyl) sulfides in 77% yield. Fluorides were trapped by trialkylsilane giving silyl fluoride. Aromatic fluorides in place of polyfluorobenzenes also reacted, provided that electron-withdrawing groups were attached. Trisulfide and tetrasulfide may also be used as the sulfur source, but they exhibit reactivities different from those of sulfur.

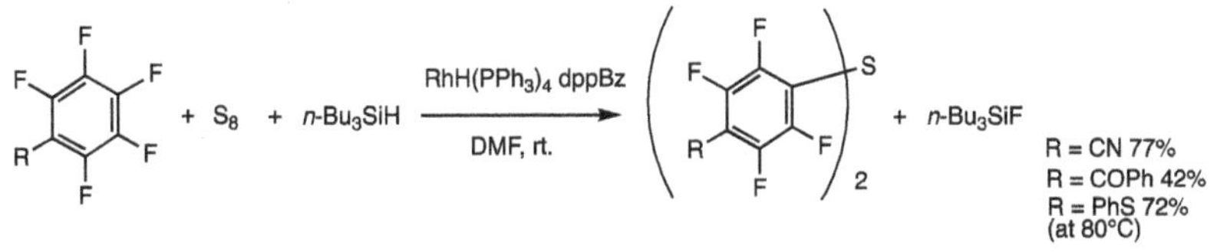

Scheme 26. Insertion reaction of sulfur and aryl fluorides.

Thiiranes are three-membered heterocyclic ring compounds containing a sulfur atom, and the insertion reaction of a sulfur atom at the alkene C=C bond is a convenient method for their synthesis [64]. The reaction may be efficiently conducted by rhodium catalysis (Scheme 27) [65]. When 7-oxabenzonorbornene, sulfur, and *p*-tolylacetylene were reacted in the presence of RhH(PPh$_3$)$_4$ (5 mol%) and dppe (10 mol%) in refluxing acetone for 3 h, exo-thiirane was obtained in 91% yield. The acetylene added stabilizes the rhodium intermediates. The isolated RhH(dppe)$_2$ also exhibited catalytic activity. The reaction is applicable to reactive alkenes, including bicyclo[2.2.1]heptenes, allenes, and (Z)-cyclooctene.

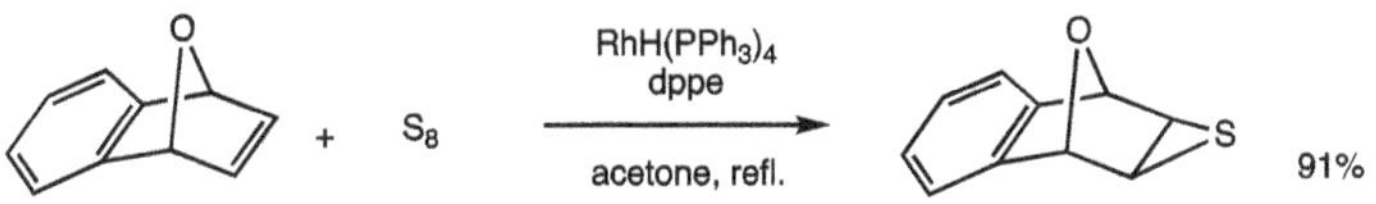

Scheme 27. Insertion reaction of sulfur and alkene.

1,4-Dithiine is a nonaromatic six-membered ring heterocyclic compound with a nonplanar structure. Dithiines are synthesized by the rhodium-catalyzed reaction of reactive alkynes and sulfur, which involves an insertion reaction of an alkyne between a S-S bond of sulfur (Scheme 28) [66]. The reaction of cyclooctyne and sulfur in refluxing 2-butanone for 3 h in the presence of RhH(PPh$_3$)$_4$ (5 mol%) and dppe (10 mol%) provided tricyclic dithiine in 53% yield. When acetylenedicarboxylate was added, unsymmetric dithiine was selectively obtained without the formation of a symmetric dithiine.

Rhodium-catalyzed synthesis of organosulfur compounds using sulfur has high potential in organic synthesis because it proceeds well below the melting point of sulfur. An interesting consideration is their application to less reactive substrates such as unstrained alkenes and alkynes, which may be developed employing thermodynamically favorable reaction systems.

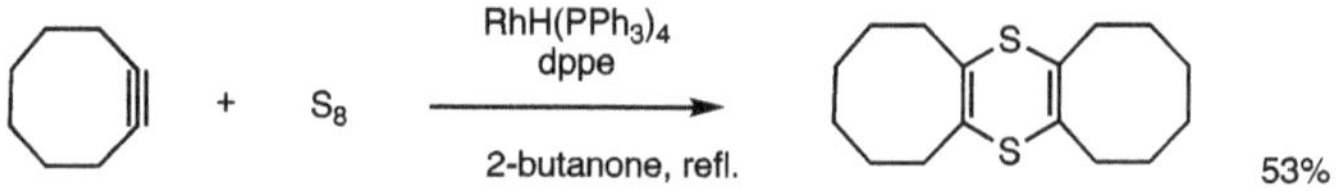

Scheme 28. Insertion reaction of sulfur and alkyne.

7. Reversible Nature of Rhodium-Catalyzed C-S Bond Formation

Rhodium-catalyzed reactions of disulfides provide various organosulfur compounds by the cleavage of RS-SR bonds and the formation of C-SR bonds, and they involve C-Rh-SR intermediates (Scheme 29). Because C-SR bond formation reactions are often reversible, C-Rh-SR intermediates can also be formed from products with C-SR bonds. The C-Rh-SR intermediates can undergo various substitution and insertion reactions to provide organic compounds. Insertion reactions of C-Rh-SR intermediates with X=Y bond compounds provide C-X-Y-SR compounds, which are generally irreversible reactions. Substitution reactions with compounds containing X-Y bonds provide novel C-X bond compounds accompanied by compounds containing RS-Y bonds, and these reactions are often reversible. This is a substitution reaction of C-SR compounds to C-X compounds, in which the role of the SR group is that of a formal leaving group. The RS-Y bond compounds are organosulfur compounds that are easy to recover and use and can be converted to other organosulfur compounds. When C-Rh-SR intermediates undergo an organothio exchange with another disulfide R'S–SR', the resulting C-Rh-SR' intermediates also undergo various reactions. This is a chemical reaction network, which provides diverse organosulfur compounds involving chemical equilibrium and interconversion reactions. In this section, we describe the reactivities of organosulfur compounds synthesized by rhodium-catalyzed reactions using disulfides.

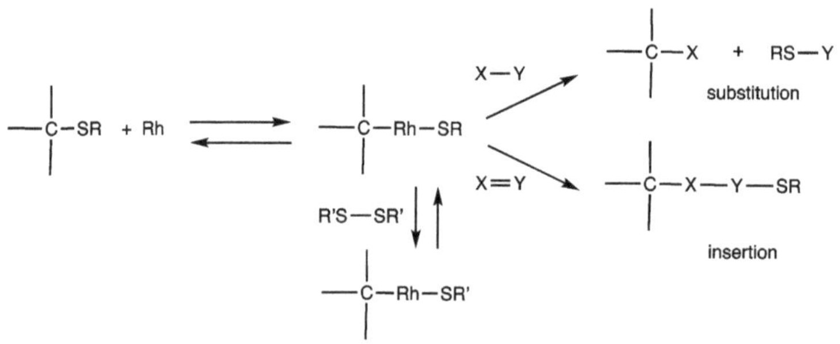

Scheme 29. Reversible nature of rhodium-catalyzed substitution reactions.

In organic synthesis, substitution reactions are generally conducted using organohalogen compounds; in this article, we describe the use of organosulfur compounds for such purpose. Halogens and organothio groups are leaving groups, which can exhibit different properties of reactions. For example, organohalogens are involved in irreversible reactions, and organosulfurs can be involved in reversible reactions, which provide diverse organosulfur compounds by slight changes in catalysts and/or reaction conditions. The organohalogen reactions form metal halides, which are not easy to recover and reuse; the organosulfur reactions provide organosulfur compounds for coproducts, which can be used for various purposes.

7.1. Reactions of 1-Thioalkynes

As shown previously, the C-S bonds in 1-thioalkynes are readily activated by rhodium catalysis (Schemes 7 and 8) [39], and then 1-thioalkynes can be used for substitution and insertion reactions. Organothio groups can be exchanged between different 1-thioalkynes, which confirmed the reversible cleavage of C-S bonds (Scheme 30) [67]. Two 1-thioalkynes in refluxing acetone in the presence of RhH(PPh$_3$)$_4$ (1 mol%) and bidentate dppf (2 mol%) provided a mixture containing comparable amounts of four possible 1-thioalkynes under chemical equilibrium. This is referred to as C-S/C-S to C-S/C-S metathesis.

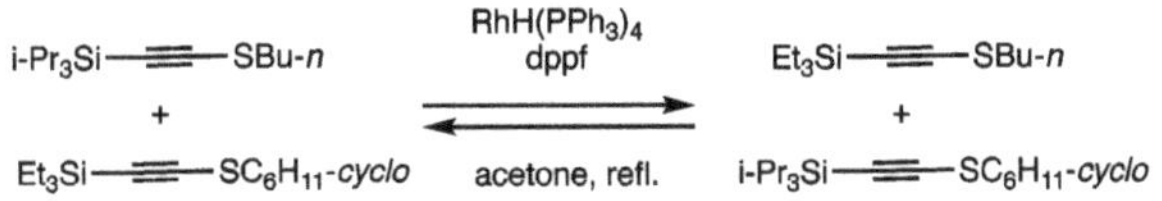

Scheme 30. Organothio exchange reaction between 1-thioalkynes.

The ligand effect was substantial in the reactions of 1-thioalkynes, and the coupling reaction of 1-thioalkynes occurred in the presence of a monodentate ligand (Scheme 31). The reaction of 1-(triethylsilyl)-2-butylthioalkyne in refluxing acetone for 2 h in the presence of RhH(PPh$_3$)$_4$ (1 mol%) and a monodentate ligand (p-MeOC$_6$H$_4$)$_3$P (3 mol%) gave 1,3-diyne in 74% yield and disulfide. This oxidative coupling reaction is referred to as C-S/C-S to C-C/S-S metathesis, which is in contrast to the above C-S/C-S to C-S/C-S metathesis (Scheme 30).

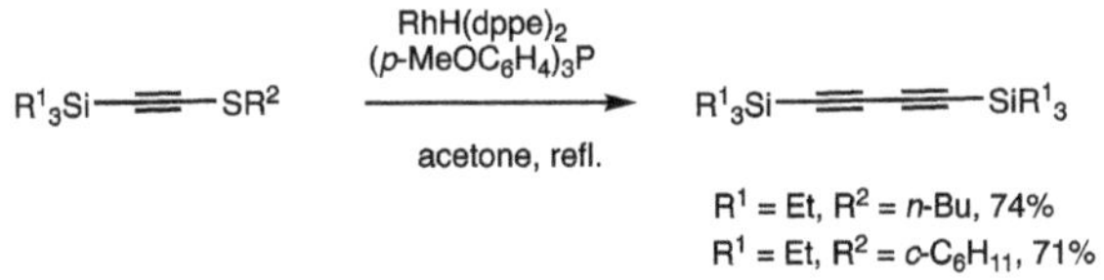

Scheme 31. Oxidative coupling reaction of 1-thioalkynes.

The insertion reaction of 1-thioalkynes and unsaturated compounds provides novel organosulfur compounds, and 1,3-butadiyne is used as a substrate (Scheme 32) [68]. The reaction of 1-butylthio-2-triethylsilyletyne and 1,4-(p-methoxyphenyl)-1,3-butadiyne in N, N-dimethylimodazolinone (DMI) at 135 °C for 6 h in the presence of RhH(PPh$_3$)$_4$ (5 mol%) and Me$_2$PhP (10 mol%) gave 1-butylthio-2-ethynyl-1,3-buten-yne in 66% yield, which is a highly unsaturated organosulfur compound.

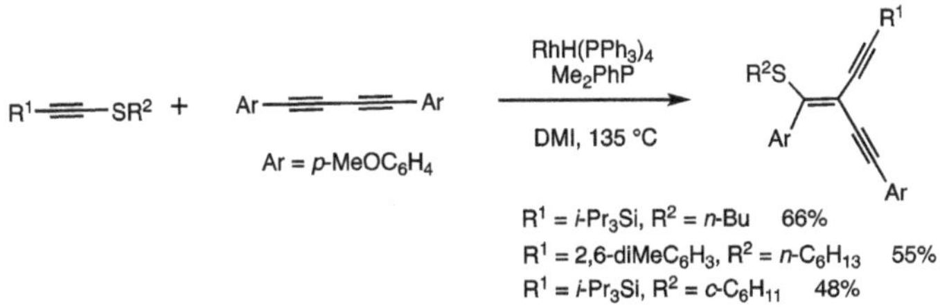

Scheme 32. Insertion reaction of 1-thioalkynes and alkynes.

The conversion of 1-thioalkynes to organophosphorus compounds proceeds under rhodium catalysis by reacting with diphosphine disulfides (Scheme 33) [53]. When 1-hexylthio-2-(2,4,6-trimethylphenyl) acetylene and tetramethyldiphosphine disulfide were reacted in refluxing chlorobenzene for 12 h in the presence of RhH(PPh$_3$)$_4$ (2 mol%) and 1,2-(diethylphosphino) ethane (depe) (4 mol%), 1-alkynylphosphine sulfide was obtained in 88% yield. 1-Thioalkynes under rhodium catalysis exhibit different reactivities from 1-alkynes and 1-haloalkynes; for example, no base or organometallic reagents are used in these reactions.

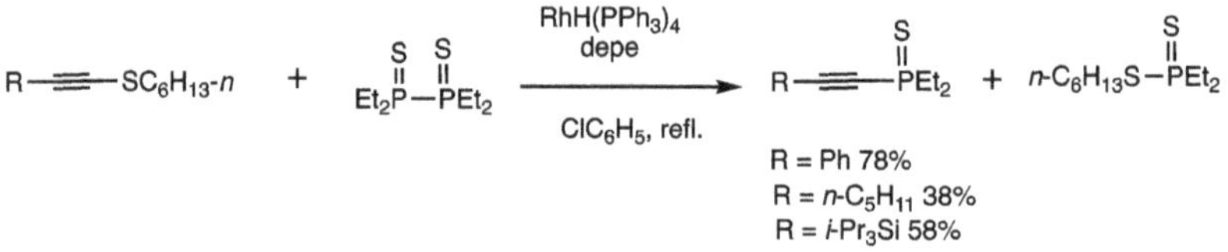

Scheme 33. Exchange reaction of diphosphine disulfide and 1-thioalkynes.

7.2. Reactions of Thioesters

Thioesters are reactive substrates in rhodium-catalyzed reactions and undergo various chemical transformations that are not observed with acyl halides. The reversible cleavage of C-S bonds was determined in studies of reactions of thioesters and disulfides (Scheme 4). Thioesters were applied to acylation reactions, forming C-S, C-P, C-N, and C-C bonds.

Thioesters react with N-organothioacylamides to provide acylimides with concomitant formation of disulfides, which is a novel acylation reaction at an amide nitrogen atom (Scheme 34) [69]. S-(p-Tolyl) benzothioate and N-(p-tolylthio)-N-butylbenzamide were reacted in refluxing chlorobenzene for 6 h in the presence of RhH(dppBz)$_2$ (5 mol%), and N-benzoyl-N-butylbenzamide was obtained in 79% yield. This is a novel metal-catalyzed coupling reaction involving N-C bond formation with cleavage of C-S and S-N bonds in two organosulfur compounds followed by oxidative elimination of disulfides.

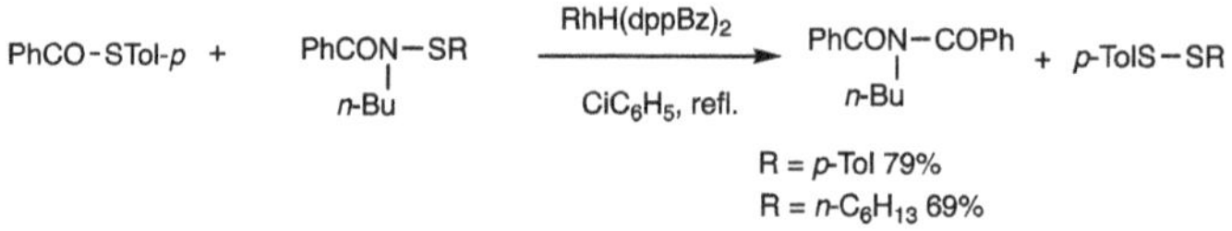

Scheme 34. Oxidative coupling reaction of thioester and N-thioamides.

Aryl methyl ethers are cleaved with thioesters under rhodium-catalyzed conditions (Scheme 35) [70]. The reaction of m-dimethoxybenzene and S-(p-tolyl) p-dimethylaminobenzothioate at 130 °C for 12 h without a solvent in the presence of RhH(CO)(PPh$_3$)$_3$ (8 mol%) and dppe (16 mol%) provided m-methoxyphenyl benzoate in 91% yield along with p-tolyl methyl sulfide. This cleavage reaction of an aryl methyl ether to provide an aryl ester proceeds under neutral conditions without using strong nucleophiles or Lewis acids.

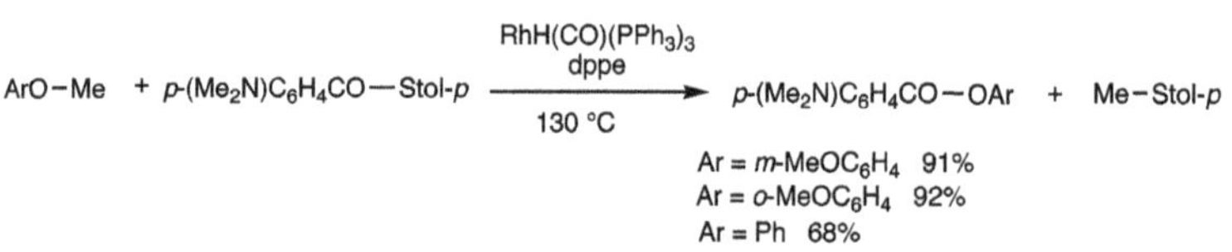

Scheme 35. Substitution reaction of thioester and aryl methyl ethers.

Heteroaryl aryl ethers are reacted with aryl thioesters to provide unsymmetric bis(heteroaryl) sulfides (Scheme 36) [71]. S-(3-Pyridyl) benzothioate and 2-benzothiazolyl 4-chlorophenyl ether were reacted in refluxing chlorobenzene for 5 h in the presence of RhH(PPh$_3$)$_4$ (5 mol%) and dppBz (10 mol%) and provided 3-pyridyl 2-benzothiazolyl sulfide in 95% yield. The other reaction pathway to form an aryl sulfide and a heteroaryl ester did not proceed. Very few unsymmetric bis(heteroaryl) sulfides were previously known, and this method provides novel multiple heteroatom compounds, some of which show interesting biological activities.

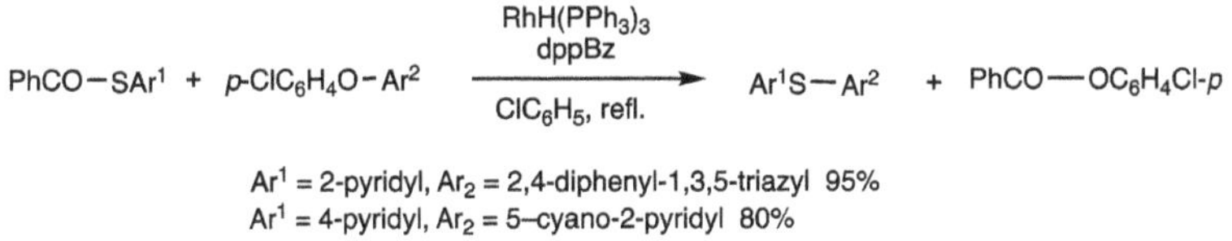
RhH(PPh3)3
dppBz
PhCO—SAr1 + p-ClC6H4O-Ar2
ClC6H5, refl.
Ar1S—Ar2 + PhCO—OC6H4Cl-p
Ar1 = 2-pyridyl, Ar2 = 2,4-diphenyl-1,3,5-triazyl 95%
Ar1 = 4-pyridyl, Ar2 = 5-cyano-2-pyridyl 80%

Scheme 36. Substitution reaction of thioester and diaryl ethers.

This reaction is applicable to the synthesis of sulfides with aliphatic cyclic groups and heteroaryl groups [72]. Specifically, the reaction of a steroidal benzothioate provided steroidal heteroaryl sulfides (Scheme 37). This is an interesting synthesis of alkyl aryl sulfides that does not require the use of bases.

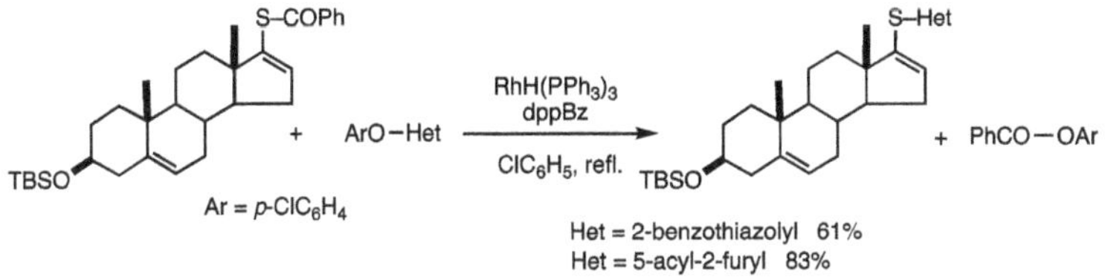
S-COPh
TBSO
+ ArO-Het
Ar = p-ClC6H4
RhH(PPh3)3
dppBz
ClC6H5, refl.
S-Het
TBSO
+ PhCO-OAr
Het = 2-benzothiazolyl 61%
Het = 5-acyl-2-furyl 83%

Scheme 37. Substitution reaction of thioester and aryl heteroaryl ethers.

Rhodium catalysts cleave C-C bonds of benzyl ketones, which then react with thioesters and esters [73]. When S-methyl benzothioate was reacted with 2-(2-thienyl)-1-(*p*-cyanophenyl)-1-ethanone in DMI at 150 °C for 12 h in the presence of $RhH(CO)(PPh_3)_3$ (10 mol%) and dppBz (20 mol%), a benzyl-exchanged ketone was obtained in 76% yield. The benzyl group was transferred between different thioesters by the cleavage of a C-C bond under chemical equilibrium. The other pathway of the bond exchange reaction providing benzyl sulfide and 1,2-diketones did not occur, which may be due to thermodynamic reasons involving the substrates and products. The use of an aryl ester in place of the thioester also gives benzyl-exchanged products.

These results indicate that the rhodium complex catalytically cleaves C-C bonds of unstrained benzyl ketones. The reaction of aryl ethers in place of thioesters/esters provided a route to form diarylmethanes [74]. The reaction of 2-(*p*-chlorophenoxy)-5-acetylfuran and 2-(1,3-benzoxazolyl)-1-phenyl-1-ethanone in refluxing chlorobenzene for 6 h in the presence of $RhH(PPh_3)_4$ (10 mol%) and dppBz (20 mol%) provided (1,3-benzoxazolyl) (5-acetylfuryl) methane in 53% yield along with phenyl benzoate. The synthesis is applicable to the production of various unsymmetric bis(heteroaryl)methanes that were previously not known. In addition, diarylmethanes are obtained from neutral compounds by the cleavage and formation of C-C bonds without using bases or organometallic reagents.

The reactions in Schemes 38 and 39 are notable examples of rhodium-catalyzed cleavage of C-C bonds in benzyl ketones, in which different types of reactions proceed depending on the substrates used (Scheme 40). The benzyl exchange reaction of esters and thioesters proceeds by CO-OAr^2 bond cleavage, which is under chemical equilibrium; the substitution reaction of ethers provided diarylmethane and esters by O-Ar^2 bond cleavage.

Rhodium-catalyzed insertion reactions of thioesters occur with 1-alkynes (Scheme 41) [49]. When *S*-butyl *p*-cyanobenzothioate was reacted with 1-decyne in DMSO at 100 °C for 12 h in the presence of $RhH(PPh_3)_4$ (5 mol%) and Et_2PhP (15 mol%), a conjugated enone was obtained in 49% yield. The benzoyl group was attached at the terminal carbon atom of the 1-alkyne and the organothio group at the internal carbon atom.

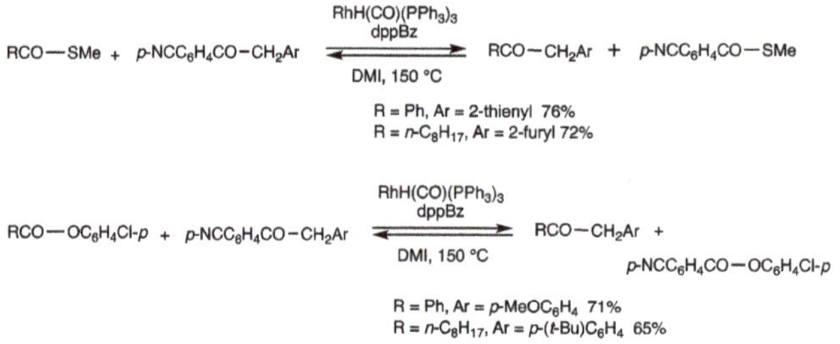

RCO—SMe + p-NCC$_6$H$_4$CO—CH$_2$Ar $\underset{\text{DMI, 150 °C}}{\overset{\text{RhH(CO)(PPh}_3)_3 \text{ dppBz}}{\rightleftharpoons}}$ RCO—CH$_2$Ar + p-NCC$_6$H$_4$CO—SMe

R = Ph, Ar = 2-thienyl 76%
R = n-C$_8$H$_{17}$, Ar = 2-furyl 72%

RCO—OC$_6$H$_4$Cl-p + p-NCC$_6$H$_4$CO—CH$_2$Ar $\underset{\text{DMI, 150 °C}}{\overset{\text{RhH(CO)(PPh}_3)_3 \text{ dppBz}}{\rightleftharpoons}}$ RCO—CH$_2$Ar + p-NCC$_6$H$_4$CO—OC$_6$H$_4$Cl-p

R = Ph, Ar = p-MeOC$_6$H$_4$ 71%
R = n-C$_8$H$_{17}$, Ar = p-(t-Bu)C$_6$H$_4$ 65%

Scheme 38. Substitution reaction of thioester/esters and arylmethyl ketones.

Ar^1—OC$_6$H$_4$Cl-p + PhCO—CH$_2$Ar^2 $\underset{\text{ClC}_6\text{H}_5\text{, refl.}}{\overset{\text{RhH(PPh}_3)_4 \text{ dppBz}}{\longrightarrow}}$ Ar^1—CH$_2$Ar^2 + PhCO—OC$_6$H$_4$Cl-p

Ar$_1$ = 5-acetyl-2-furyl, Ar$_2$ = 1,3-benzothiazolyl 53%
Ar$_1$ = 4,5-diphenyl-1,3-oxazolyl, Ar$_2$ = 4-pyridyl 72%
Ar$_1$ = 6,7-dimethoxy-1,3-naphthyridyl, Ar$_2$ = 2-thienyl 64%

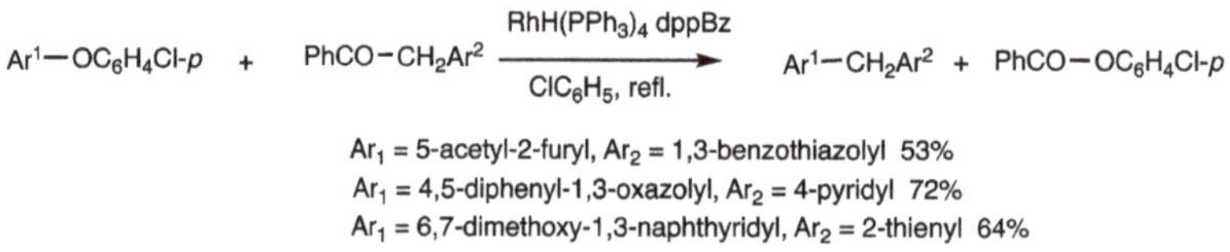

Scheme 39. Substitution reaction of diaryl ethers and arylmethyl ketones.

R^{1}CO—CH$_2$Ar^1 + R^{2}CO—OAr^2 $\rightleftharpoons$ R^{2}CO—CH$_2$Ar^1 + R^{1}CO—OAr^2

R^{1}CO—CH$_2$Ar^1 + Ar^2—OR^3 $\longrightarrow$ Ar^2—CH$_2$Ar^1 + R^{1}CO—OR^3

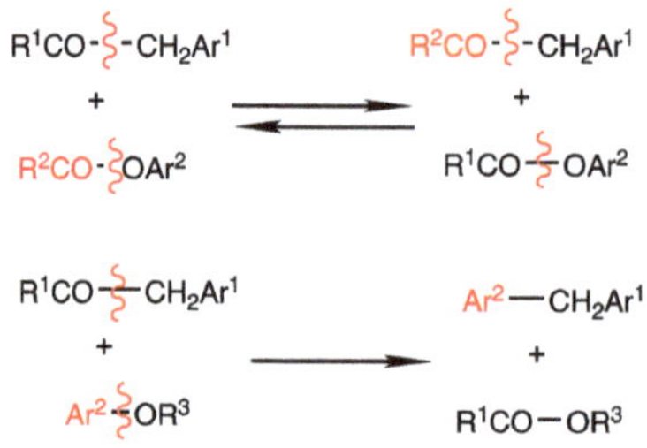

Scheme 40. Comparison of substitution reactions of arylmethyl ketones and aryl esters/aryl ethers.

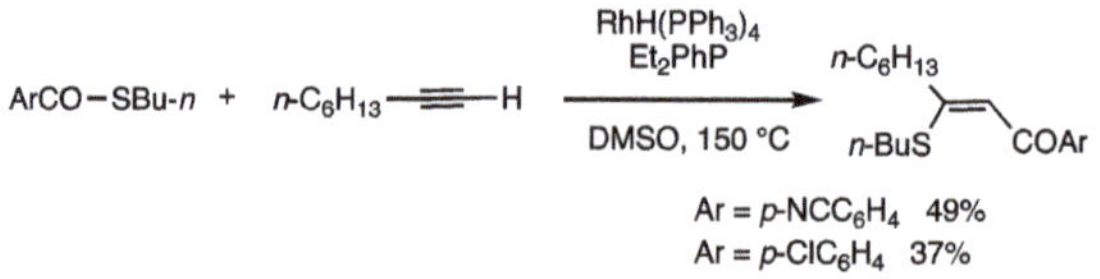

ArCO—SBu-n + n-C$_6$H$_{13}$—≡—H $\underset{\text{DMSO, 150 °C}}{\overset{\text{RhH(PPh}_3)_4 \text{ Et}_2\text{PhP}}{\longrightarrow}}$ n-C$_6$H$_{13}$\\=/COAr (n-BuS)

Ar = p-NCC$_6$H$_4$ 49%
Ar = p-ClC$_6$H$_4$ 37%

Scheme 41. Insertion reaction of thioesters and 1-alkyne.

Insertion reactions of thioesters occur with strained alkenes (Scheme 42) [75]. The reaction of S-(p-tolyl) benzothioate and norbornadiene under THF reflux for 6 h in the presence of Pd$_2$(dba)$_3$ (dba = dibenzylideneacetone) (5 mol%) and (2,4,6-(MeO)$_3$C$_6$H$_2$)$_3$P (20 mol%) gave the acylated adduct with trans configuration, in which an acyl group is attached in the endo position. The initial bond formation is considered to involve acylation, which is followed by thiolation.

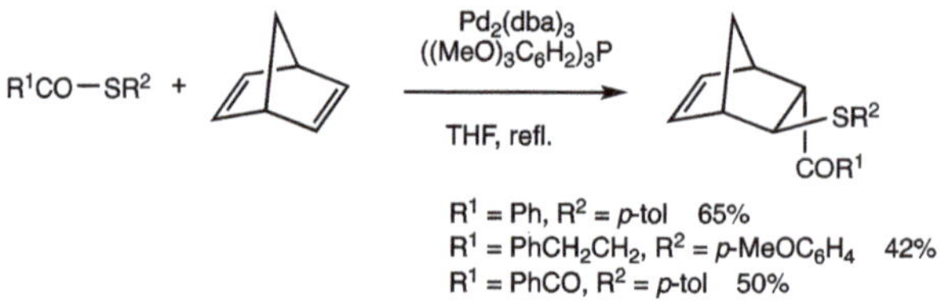

R^{1}CO—SR^2 + [norbornadiene] $\underset{\text{THF, refl.}}{\overset{\text{Pd}_2(\text{dba})_3 \text{ ((MeO)}_3\text{C}_6\text{H}_2)_3\text{P}}{\longrightarrow}}$ [norbornene-SR^2/COR^1]

R^1 = Ph, R^2 = p-tol 65%
R^1 = PhCH$_2$CH$_2$, R^2 = p-MeOC$_6$H$_4$ 42%
R^1 = PhCO, R^2 = p-tol 50%

Scheme 42. Insertion reaction of thioesters and alkene.

7.3. Reactions of α-Thioketone

The C-S bonds in α-thioketones are activated by rhodium catalysts to form oxidative addition intermediates with S-Rh-C subunits, which react with C-H bonds in organic compounds (Scheme 11) [43,44]. The activation of α-thioketones can be used for the coupling reaction to form C-C bonds at the ketone α-position (Scheme 43) [76]. The reaction of 1,2-diphenyl-1-ethanone and methylthiomethyl *t*-butyl ketone in refluxing chlorobenzene for 6 h in the presence of RhH(PPh$_3$)$_4$ (10 mol%) and dppBz (20 mol%) gave a dimeric ketone at the α-position in 67% yield. The reaction is considered to involve the initial transfer of the methylthio group to 1,2-diphenyl-1-ethanone followed by coupling to form diketone along with dimethyl disulfide. The mechanism was determined from the reaction between α-methylthiolated 1,2-diphenyl-1-ethanone and 1,2-diphenyl-1-ethanone to provide the coupling product. It should be noted that oxidative dimerization of ketones can be conducted under rhodium catalysis using α-thioketones as oxidation reagents.

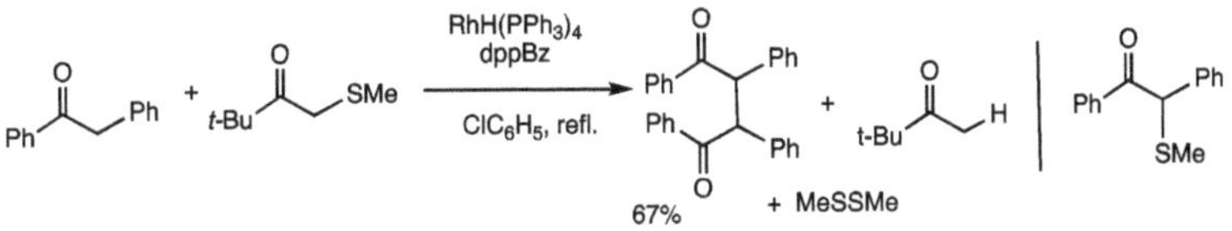

Scheme 43. Oxidative coupling reaction of ketones.

7.4. Reactions of Aryl/Heteroaryl Sulfides

The rhodium-catalyzed C-S bond cleavage of aryl sulfides provided other aryl sulfides by the exchange of arylthio groups [77]. Polyfluorophenyl and heteroaryl derivatives are highly reactive substrates for the cleavage and formation of C-S bonds, and symmetric diaryl sulfides can be converted into unsymmetric diaryl sulfides (Scheme 44). Perfluorinated bis(*p*-benzoylphenyl) sulfide, (phenylthiol)pentafluorobenzene, and triisopropylsilane were reacted in the presence of RhH(PPh$_3$)$_4$ (5 mol%) and dppBz (10 mol%) in refluxing THF for 6 h and provided unsymmetric diaryl sulfide in 51% yield. Silane was added to trap fluoride by the formation of silyl fluoride, which resulted in an exothermic reaction. The reaction is under chemical equilibrium without silane, and the unsymmetric diaryl sulfide can also undergo the aryl exchange reaction. Combined with the synthesis of symmetric diaryl sulfides from polyfluorobenzene and sulfur (Scheme 26), various unsymmetric diaryl sulfides can be synthesized using sulfur.

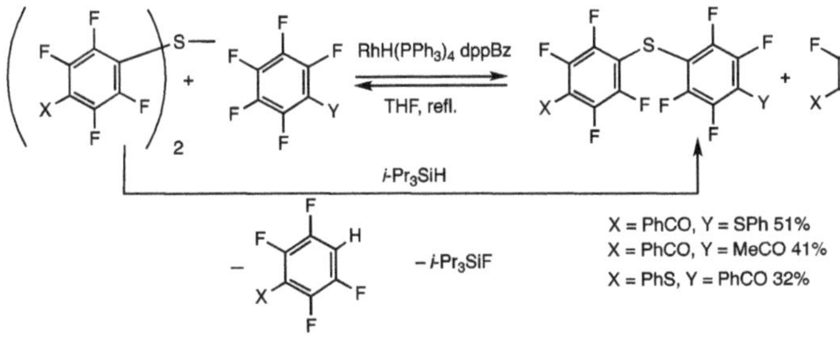

X = PhCO, Y = SPh 51%
X = PhCO, Y = MeCO 41%
X = PhS, Y = PhCO 32%

Scheme 44. Exchange reaction of diaryl sulfides and aryl fluorides.

Benzo-fused heteroarenes are thiolated under rhodium-catalyzed conditions using α-phenylthioisobutyrophenone (Scheme 11), and nonbenzo-fused heteroarenes are thiolated using 2-methylthiothiazole (Scheme 45) [78]. The former thiolating reagent is more reactive than disulfides and α-thioketones because the less acidic nature of the 2-proton of 2-(methylthio)thiazole causes the chemical equilibrium to favor the formation of thiolated nonbenzo-fused heteroarenes.

The reaction of 3-phenylthiazole and 2-(methylthio) benzothiazole (3 equivalents) in the presence of RhH(PPh$_3$)$_4$ (10 mol%) and 1,3-(dicyclohexylphosphino)-1,3-propane (dcypp) (20 mol%) in refluxing 1,2-dichlorobenzene for 3 h gave 2-methylthio-3-phenylthiazole in 74% yield. This reaction is under chemical equilibrium, and removal of volatile thiazole increased the yield. The reaction is applicable to various substituted thiazoles and oxazoles.

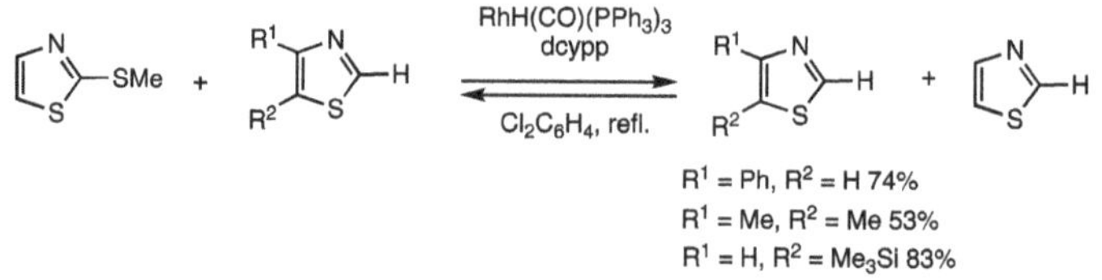

Scheme 45. Organothio exchange reaction of thiazoles.

Rhodium-catalyzed organothiolation reactions of heteroarenes and related compounds using disulfides [79–89] and sulfur [90] have recently been reported. These methods employ stoichiometric or substoichiometric amounts of copper(II) or silver(I) salts. The strong metal oxidizing reagents are considered to regenerate higher oxidation states of rhodium complexes and produce copper(I) salts or copper/silver metals as byproducts. The present reaction is characterized by simple transfer of organothio groups without using such metal reagents, which is achieved by judicious choice of organothio donor. A strong metal oxidizing reagent is an equivalent of a large amount of energy, and a related discussion will be provided in Section 8 on the use of the strong metal base of sodium hydroxide (Scheme 47).

2,5-Disubstituted 1,4-dithiines undergo rearrangement to provide 2,6-disubstituted derivatives, which involves the cleavage of two C-S bonds (Scheme 46) [91]. The reaction requires rhodium catalysis and is under chemical equilibrium. When 2,5-di(*t*-butyl)-1,4-dithiine was treated with RhH(dppe)$_2$ (10 mol%) and dimethyl acetylenedicarboxylate (DMAD) (3 equivalents) in refluxing toluene for 24 h, 1,6-di(*t*-butyl)-1,4-dithiine was obtained in 48% yield, along with the starting material in 51% recovery yield. The role of acetylene was to stabilize the intermediate rhodium complex. Under forced conditions of 150 °C without a solvent, one of the olefin moieties in the starting material is exchanged with DMAD. The formation of a rhodacycle intermediate and the involvement of the thio-Diels–Alder reaction are proposed to occur.

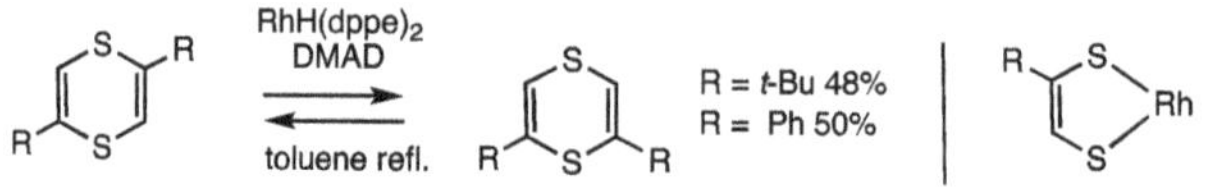

Scheme 46. Exchange reaction of 1,4-dithiines.

Organosulfur compounds can be transformed into other organosulfur compounds and related organic compounds by the rhodium-catalyzed method owing to the reversible nature of the reactions. Such manipulation provides a diversity of derivatives starting from a single organosulfur compound by changing their cosubstrates.

8. Conclusions

8.1. Mechanisms of Rhodium-Catalyzed Substitution and Insertion Reactions of Disulfides

Rhodium complexes catalyze the cleavage of disulfide S-S bonds and transfer of organothio groups to other organic compounds, which provides various organosulfur compounds. Mechanistic models of substitution reactions and insertion reactions are shown below (Figure 7). A low-valent Rh(I) complex undergoes oxidative addition with a disulfide to provide a S-Rh(III)-S complex (Figure 7a).

Such oxidative addition reactions of disulfides are known [92–94]. Formation of dithiorhodacycles by reactions of rhodium complexes and sulfur has also been reported [65,95]. Then, ligand exchange proceeds with a molecule possessing an X-Y bond to form an X-Rh(III)-S complex, and reductive elimination provides a product possessing a S-X bond with the regeneration of the Rh(I) complex. Alternatively, formation of a Rh(V) complex can be considered by subsequent oxidative addition of the S-Rh(III)-S complex with a molecule possessing an X-Y bond.

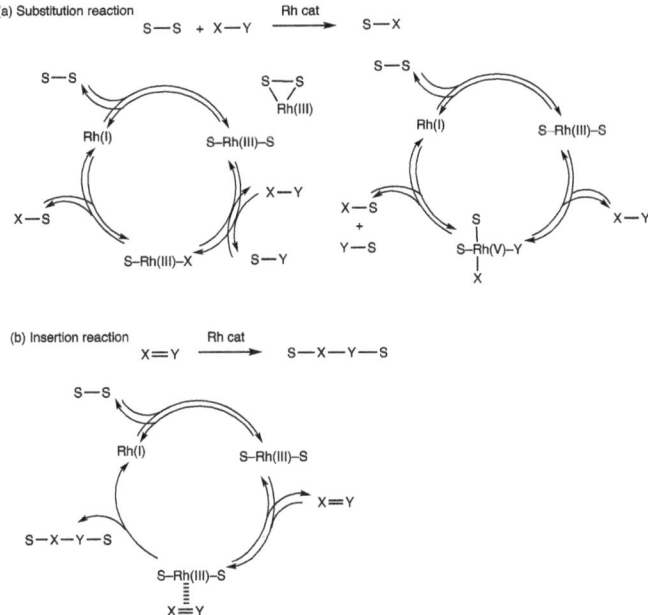

Figure 7. Mechanistic models of rhodium-catalyzed (**a**) substitution reactions and (**b**) insertion reactions with disulfides.

Two reductive eliminations then provide two molecules containing X-S and Y-S bonds, which are accompanied by regeneration of the Rh(I) complex. Formation of Rh(V) complexes has been reported [96–99]. A mechanistic model of the insertion reaction can also involve oxidative addition of a Rh(I) complex and a disulfide, which is followed by the transfer of two organothio groups to an unsaturated molecule possessing an X=Y bond to form a product possessing a S-X-Y-S group with the regeneration of the Rh(I) complex (Figure 7b).

Involvement of chemical equilibria is a notable feature of the present rhodium-catalyzed substitution reactions using disulfides. The above mechanistic model shows reversible nature in the oxidative addition of a rhodium(I) complex and a disulfide (Figure 7a). In addition, ligand exchange of a S-Rh(III)-S complex possessing an X-Y molecule is reversible; reductive elimination of an X-Rh(III)-S complex to provide a product with a S-X bond may be reversible. Such a sequence of reversible reactions involving organosulfur–rhodium complexes provides a chemical equilibrium. Disulfide exchange reactions described in Section 2 support the reversible mechanistic model.

Insertion reactions can also involve reversible oxidative addition of a Rh(I) complex and a disulfide (Figure 7b). Subsequent insertion of an X=Y molecule may be irreversible because reactions of unsaturated compounds to form saturated compounds are generally exothermic.

It should be noted that the formation of Rh-S bonds can be reversible, which may be a basis for the rhodium-catalyzed synthesis of organosulfur compounds described in this article. This is in contrast to a general thought that chemical bonds between transition metals and sulfur atoms are very strong and make such catalysis difficult.

8.2. Rhodium-Catalyzed Synthesis of Diverse Organosulfur Compounds

The rhodium-catalyzed synthetic method developed in this study provides organosulfur compounds with diverse structures; the development of highly active catalysts and design of exothermic reactions have been critical for this method. The synthesis employs disulfides and sulfur possessing S-S bonds, which are stable and readily available. Rhodium complexes cleave S-S bonds and promote substitution and insertion reactions with various organic compounds.

None of these reactions employ bases or organometallic reagent. This feature is in contrast to conventional syntheses of organosulfur compounds using thiolate anions and organohalogen compounds, which rely on the exothermic nature of reactions owing to the neutralization of hydrogen halides. Such a reaction involving rhodium catalysis was also reported [100]. In this context, the role of a base is considered, for example, in a reaction generating hydrogen chloride, which is a substitution reaction of an alkyl chloride and a thiol. Neutralization of hydrogen chloride with sodium hydroxide to form sodium chloride is strongly exothermic, $\Delta H = -285.8$ kJ mol^{-1}, as calculated from the heat of formation (Scheme 47). The regeneration of sodium hydroxide from sodium chloride by electrolysis requires a large amount of energy: $\Delta H = +271.4$ kJ mol^{-1}. No base is incorporated in the product, and the neutralization–regeneration cycle consumes energy to make the reaction exothermic. Metal halides are not easy to recover and reuse, and to do so requires a large amount of energy. A strong base is an equivalent of a large amount of energy; for example, the production of sodium hydroxide from sodium chloride (1×10^{10} kWh in 2017) [101] consumes 1% of all electricity in Japan (8×10^{11} kWh in 2015) [102]. The atom economy is often employed to analyze the efficiency of a synthetic chemical reaction [103]. The formation of metal halides can provide a favorable atom economy because their molecular weights are small. However, it is also critical to consider energy, and the reuse of metal halides requires a vast amount of energy.

$$NaOH\ (c) + HCl\ (g) \longrightarrow NaCl\ (c) + H_2O\ (l)$$
$$\Delta H = -285.8 \text{ kJ mol}^{-1}$$

$$NaCl\ (c) + H_2O\ (l) \longrightarrow NaOH\ (c) + 1/2 H_2O\ (l) + 1/2 Cl_2\ (g)$$
$$\Delta H = +271.4 \text{ kJ mol}^{-1}$$

Scheme 47. Thermodynamic analysis of the use of sodium hydroxide.

Another notable aspect of the syntheses presented herein is the provision of diverse organic compounds that are structurally related. This feature is derived from the use of disulfides (RS-SR) and sequential substitution reactions (Figure 8). When a rhodium-catalyzed reaction converts one SR group with **A**, a series of products RS-**A**^{1}, RS-**A**^{2}, RS-**A**^{3}, RS-**A**^{4}, ... are produced. Then, the rhodium-catalyzed organothio exchange reaction of RS-**A**^{1} provides diverse organosulfur compounds R^{1}S-**A**^{1}, R^{2}S-**A**^{1}, R^{3}S-**A**^{1}, ... , as described in Sections 2, 3 and 6. Using the reversible nature of the rhodium-catalyzed reactions, we can substitute the RS group in RS-**A**^{1} with **B**^{1}, which provides coupling products **B**^{1}-**A**^{1}, **B**^{1}-**A**^{2}, **B**^{1}-**A**^{3}, ... Rhodium catalysis then provides other coupling products **B**^{2}-**A**^{1}, **B**^{2}-**A**^{2}, **B**^{2}-**A**^{3}, ... , as described in Section 7. Thus, organothio groups derived from disulfides RS-SR have dual roles. One is as a part of organosulfur compounds R^{n}S-**A**^{m}; the other is as a leaving group to provide coupling products **B**^{n}-**A**^{m}. This chemistry provides a systematic method to construct a library of compounds in an energy-saving manner.

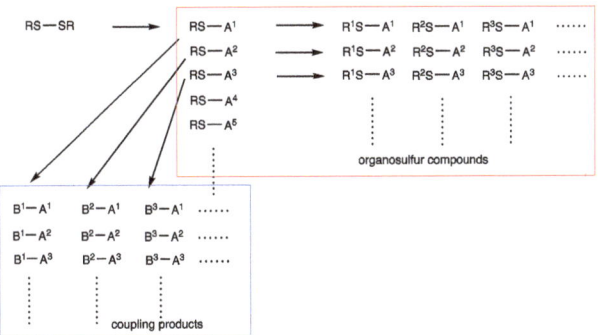

Figure 8. Synthesis of diverse organosulfur compounds and coupling products by rhodium-catalyzed reactions of disulfides.

8.3. Chemical Reaction Network System

The rhodium-catalyzed synthesis of organosulfur compounds involves reversible reactions, which are chemical equilibria and interconversion reactions (Figure 9). Chemical equilibrium in closed systems involves a comparable thermodynamic stability of substrates and products, along with a low energy barrier (Figure 4). Such reactions are shown by straight arrows pointing in opposite directions. Interconversion reactions in open systems involve the addition of appropriate organic cosubstrates and the formation of coproducts, which inverts the relative thermodynamic stability of substrates and products. The reactions can then proceed in either direction, as shown by curved arrows pointing in opposite directions. The summary of this work provides a chemical reaction network with many reversible reactions, in which substrates and products are mutually related (Figure 9). The product of a reaction can be a substrate in the next or distantly located chemical reactions. This chemical reaction network, however, is a simplified model based on the reactions of disulfide and sulfur, and the real network is multidimensional. As examples of such networks, we have previously proposed an acylation reaction network [49] and a C-H thiolation reaction network [29].

Biological systems employ complex chemical reaction networks with energy-saving characteristics. Chemical equilibria can be controlled by flow systems in biological cells, and interconversion reactions can be controlled by external energy using ATP and ADP to produce exothermic reactions. Biological reaction networks are further controlled by enzymatic catalysis [6,7]. The chemical reaction network herein can be a model of biological systems involving many chemical equilibria and interconversion reactions, although biological systems are much more well-organized and functional.

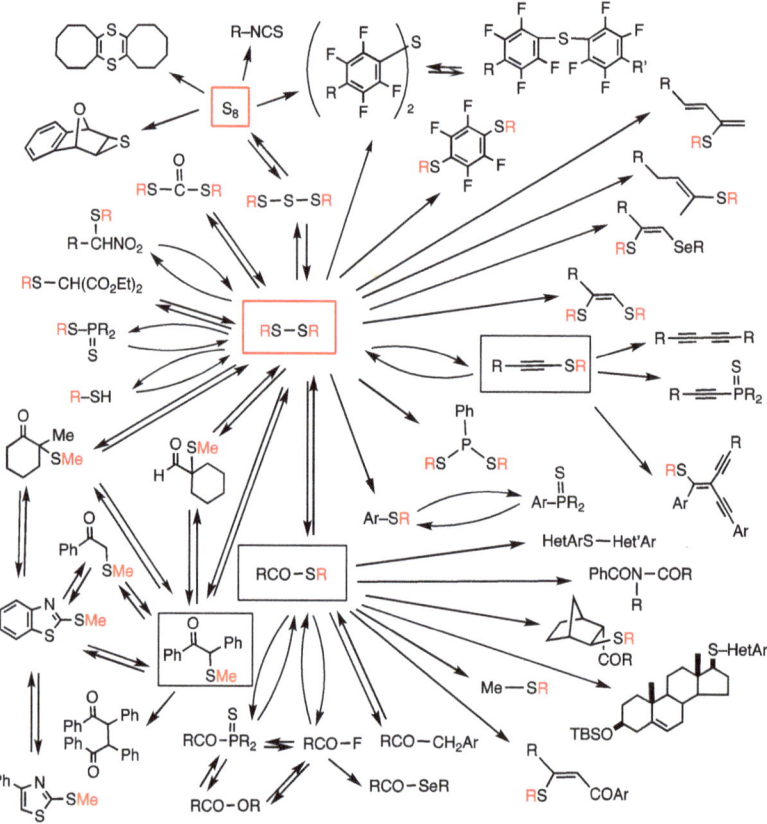

Figure 9. Schematic presentation of a rhodium-catalyzed chemical reaction network using disulfide and sulfur. Straight arrows pointing in opposite directions show chemical equilibria in closed systems, and curved arrows pointing in opposite directions show interconversion reactions in open systems. Red R letters are derived from organic groups in disulfides. Squares contain important synthetic intermediates.

Author Contributions: Writing, M.A. and M.Y. All authors have read and agreed to the published version of the manuscript.

Funding: This research was supported by the Japan Agency for Medical Research and Development (AMED), Platform Project for Supporting Drug Discovery and Life Science Research (Grant Number JP19 am0101100), Nagase Science and Technology Foundation, Kobayashi Foundation, and Tohoku University Center for Gender Equality Promotion (TUMUG).

Conflicts of Interest: The authors declare no conflict of interest.

References

1. Kennepohl, D.; Farmer, S.; Reusch, W.; Soderberg, T.; Schaller, C.P. Available online: https://chem.libretexts.org/Bookshelves/Organic_Chemistry/Supplemental_Modules_(Organic_Chemistry)/Thiols_and_Sulfides/Thiols_and_Sulfides (accessed on 5 August 2020).
2. Schaeffer, C.D., Jr.; Strausser, C.A.; Thomsen, M.W.; Yoder, C.H. Data for General, Organic, and Physical Chemistry. Available online: http://chembook.weebly.com/uploads/2/5/7/7/257728/schaeffer_cd_data_for_general_organic_and_physical_chemistry.pdf (accessed on 5 August 2020).

3. Benson, S.W. Thermochemistry and Kinetics of Sulfur-Containing Molecules and Radicals. *Chem. Rev.* **1978**, *78*, 23–35. [CrossRef]
4. Koval', I.V. The Chemistry of Disulfides. *Russ. Chem. Rev.* **1994**, *63*, 735–750.
5. Voronkov, M.G.; Deryagina, E.N. Thermal Reactions of Thiyl Radicals. *Russ. Chem. Rev.* **1990**, *59*, 778–791. [CrossRef]
6. Alberts, B.; Johnson, A.; Lewis, J.; Morgan, D.; Raff, M.; Robert, K.; Walter, P. *Molecular Biology of the Cell*, 6th ed.; Garland Science: New York, NY, USA, 2014.
7. Lodish, H.; Berk, A.; Kaiser, C.A.; Krieger, M.; Scott, M.P.; Bretscher, A.; Ploegh, H.; Matsudaira, P. *Molecular Cell Biology*, 6th ed.; W.H. Freeman and Company: New York, NY, USA, 2008.
8. Fass, D.; Thorpe, C. Chemistry and Enzymology of Disulfide Cross-Linking in Proteins. *Chem. Rev.* **2018**, *118*, 1169–1198. [CrossRef]
9. Raina, S.; Missiakas, D. Making and Breaking Disulfide Bonds. *Ann. Rev. Microbiol.* **1997**, *51*, 179–202. [CrossRef]
10. Chiu, J.; Hogg, P.J. Allosteric Disulfides: Sophisticated Molecular Structures Enabling Flexible Protein Regulation. *J. Biol. Chem.* **2019**, *294*, 2949–2960. [CrossRef]
11. Hogg, P.J. Disulfide Bonds as Switches for Protein Function. *Trends Biochem. Sci.* **2003**, *28*, 210–214. [CrossRef]
12. Wang, N.; Saidhareddy, P.; Jiang, X. Construction of Sulfur-containing Moieties in the Total Synthesis of Natural Products. *Nat. Prod. Rep.* **2020**, *37*, 246–275. [CrossRef]
13. Zhang, Y.; Glass, R.S.; Char, K.; Pyun, J. Recent Advances in the Polymerization of Elemental Sulphur, Inverse Vulcanization and Methods to Obtain Functional Chalcogenide Hybrid Inorganic/Organic Polymers (CHIPs). *Polym. Chem.* **2019**, *10*, 4078–4105.
14. Bang, E.-K.; Lista, M.; Sforazzini, G.; Sakai, N.; Matile, S. Poly(disulfide)s. *Chem. Sci.* **2012**, *3*, 1752–1763. [CrossRef]
15. Mandal, B.; Basu, B. Recent Advances in S–S Bond Formation. *RSC Adv.* **2014**, *4*, 1385–13881. [CrossRef]
16. Witt, D. Recent Developments in Disulfide Bond Formation. *Synthesis* **2008**, 2491–2509. [CrossRef]
17. Carey, F.; Sundberg, R.J. *Advanced Organic Chemistry: Part A: Structure and Mechanisms*, 15th ed.; Springer: Berlin, Germany, 2007.
18. Otocka, S.; Kwiatkowska, M.; Madalińsska, L.; Kiełbasiński, P. Chiral Organosulfur Ligands/Catalysts with a Stereogenic Sulfur Atom: Applications in Asymmetric Synthesis. *Chem. Rev.* **2017**, *117*, 4147–4181. [CrossRef] [PubMed]
19. Kaiser, D.; Klose, I.; Oost, R.; Neuhaus, J.; Maulide, N. Bond-Forming and -Breaking Reactions at Sulfur(IV): Sulfoxides, Sulfonium Salts, Sulfur Ylides, and Sulfinate Salts. *Chem. Rev.* **2019**, *119*, 8701–8780. [CrossRef]
20. Yang, S.; Feng, B.; Yang, Y. Rh(III)-Catalyzed Direct *ortho*-Chalcogenation of Phenols and Anilines. *J. Org. Chem.* **2017**, *82*, 12430–12438. [CrossRef]
21. Mukaiyama, T. Oxidation-Reduction Condensation. *Angew. Chem. Int. Ed.* **1976**, *15*, 94–103. [CrossRef]
22. Kuniyasu, H.; Ogawa, A.; Miyazaki, S.; Ryu, I.; Kambe, N.; Sonoda, N. Palladium-Catalyzed Addition and Carbonylative Addition of Diaryl Disulfides and Diselenides to Terminal Acetylenes. *J. Am. Chem. Soc.* **1991**, *113*, 9796–9803. [CrossRef]
23. Beletskaya, I.P.; Ananikov, V.P. Transition-Metal-Catalyzed C−S, C−Se, and C−Te Bond Formation via Cross-Coupling and Atom-Economic Addition Reactions. *Chem. Rev.* **2011**, *111*, 1596–1636. [CrossRef]
24. Kondo, T.; Uenoyama, S.; Fujita, K.; Mitsudo, T. First Transition-Metal Complex Catalyzed Addition of Organic Disulfides to Alkenes Enables the Rapid Synthesis of *Vicinal*-Dithioethers. *J. Am. Chem. Soc.* **1999**, *121*, 482–483. [CrossRef]
25. Qi, X.; Li, Y.; Bai, R.; Lan, Y. Mechanism of Rhodium-Catalyzed C−H Functionalization: Advances in Theoretical Investigation. *Acc. Chem. Res.* **2017**, *50*, 2799–2808. [CrossRef]
26. Arisawa, M.; Tanii, S.; Tazawa, T.; Yamaguchi, M. Synthesis of Unsymmetric HetAr–X–HetAr′ Compounds by Rhodium-Catalyzed Heteroaryl Exchange Reactions. *Heterocycles* **2017**, *94*, 2179–2207. [CrossRef]
27. Arisawa, M. Rhodium-catalyzed Synthesis of Unsymmetric Di(heteroaryl) Compounds via Heteroaryl Exchange Reactions. *Phosphorus Sulfur Silicon Rel. Elem.* **2019**, *194*, 643–648. [CrossRef]
28. Arisawa, M.; Yamaguchi, M. Transition-metal-catalyzed Synthesis of Organosulfur Compounds. *Pure Appl. Chem.* **2008**, *80*, 993–1003. [CrossRef]

29. Arisawa, M. Synthesis of Organosulfides Using Transition-metal-catalyzed Substitution Reactions: To Construct Exergonic Reactions Employing Metal Inorganic and Organic Co-substrate/co-product Methods. *Tetrahedron Lett.* **2014**, *55*, 3391–3399. [CrossRef]
30. Arisawa, M.; Yamaguchi, M. Rhodium-Catalyzed Synthesis of Organosulfur Compounds using Sulfur. *Synlett* **2019**, *30*, 1621–1631.
31. Arisawa, M. Transition-metal-catalyzed Synthesis of Organophosphorus Compounds Involving P-P Bond Cleavage. *Synthesis* **2020**. [CrossRef]
32. Arisawa, M.; Yamaguchi, M. Rhodium-Catalyzed Disulfide Exchange Reaction. *J. Am. Chem. Soc.* **2003**, *125*, 6624–6625. [CrossRef]
33. Miyagawa, M.; Arisawa, M.; Yamaguchi, M. Equilibrium Shift Induced by Chiral Nanoparticle Precipitation in Rhodium-Catalized Disulfide Exchange Reaction. *Tetrahedron* **2015**, *71*, 4920–4926. [CrossRef]
34. Arisawa, M.; Suwa, A.; Yamaguchi, M. $RhCl_3$-catalyzed Disulfide Exchange Reaction Using Water Solvent in Homogeneous and Heterogeneous Systems. *J. Organomet. Chem.* **2006**, *691*, 1159–1168. [CrossRef]
35. Arisawa, M.; Kuwajima, M.; Suwa, A.; Yamaguchi, M. $RhCl_3$-Catalyzed Disulfide Exchange Reaction of Insulfin and Dithiodiglycolic Acid. *Heterocycles* **2010**, *80*, 1239–1248.
36. Arisawa, M.; Kubota, T.; Yamaguchi, M. Rhodium-Catalyzed Alkylthio Exchange Reaction of Thioester and Disulfide. *Tetrahedron Lett.* **2008**, *49*, 1975–1978. [CrossRef]
37. Arisawa, M.; Yamada, T.; Yamaguchi, M. Rhodium-Catalyzed Interconversion between Acid Fluorides and Thioesters Controlled using Heteroatom Acceptors. *Tetrahedron Lett.* **2010**, *51*, 6090–6092. [CrossRef]
38. Arisawa, M.; Suzuki, R.; Ohashi, K.; Yamaguchi, M. Rhodium-Catalyzed Synthesis of Heteroarylselenyl Esters from Diheteroaryl Diselenides and Acid Fluorides. *Asian J. Org. Chem.* **2020**, *9*, 553–556. [CrossRef]
39. Arisawa, M.; Fujimoto, K.; Morinaka, S.; Yamaguchi, M. Equilibrating C-S Bond Formation by C-H and S-S Bond Metathesis. Rhodium-Catalyzed Alkylthiolation Reaction of 1-Alkynes with Disulfides. *J. Am. Chem. Soc.* **2005**, *127*, 12226–12227. [CrossRef] [PubMed]
40. Arisawa, M.; Nihei, Y.; Yamaguchi, M. Rhodium-Catalyzed Arylthiolation Reaction of Nitroalkanes, Diethyl Malonate, and 1,2-Diphenylethanone with Diaryl Disulfides: Control of Disfavored Equilibrium Reaction. *Tetrahedron Lett.* **2012**, *53*, 5729–5732. [CrossRef]
41. Arisawa, M.; Toriyama, F.; Yamaguchi, M. Rhodium-Catalyzed Organothio Exchange Reaction of α-Organothioketones with Disulfides. *Chem. Pharm. Bull.* **2010**, *58*, 1349–1352. [CrossRef]
42. Arisawa, M.; Toriyama, F.; Yamaguchi, M. An Activated Catalyst $RhH(PPh_3)_4$-dppe-Me_2S_2 for α-Methylthiolation of α-Phenyl Ketones. *Heteroatom Chem.* **2011**, *22*, 18–23. [CrossRef]
43. Arisawa, M.; Suwa, K.; Yamaguchi, M. Rhodium-Catalyzed Methylthio Transfer Reaction between Ketone α-Positions: Reversible Single-Bond Metathesis of C-S and C-H Bonds. *Org. Lett.* **2009**, *11*, 625–627. [CrossRef]
44. Arisawa, M.; Toriyama, F.; Yamaguchi, M. Rhodium-catalyzed α-Methylthiolation Reaction of Unactivated Ketones Using 1,2-Diphenyl−2-Methylthio−1-Ethanone for the Methylthio Transfer Reagent. *Tetrahedron* **2011**, *67*, 2305–2312. [CrossRef]
45. Arisawa, M.; Toriyama, F.; Yamaguchi, M. Rhodium-catalyzed Phenylthiolation Reaction of Heteroaromatic Compounds using α-(Phenylthio) isobutyrophenone. *Tetrahedron Lett.* **2011**, *52*, 2344–2347. [CrossRef]
46. Macgregor, S.A.; Roe, D.C.; Marshall, W.J.; Bloch, K.M.; Bakhmutov, V.I.; Grushin, V.V. The F/Ph Rearrangement Reaction of $[(Ph_3P)_3RhF]$, the Fluoride Congener of Wilkinson's Catalyst. *J. Am. Chem. Soc.* **2005**, *127*, 15304–15321. [CrossRef] [PubMed]
47. Arisawa, M.; Suzuki, T.; Ishikawa, T.; Yamaguchi, M. Rhodium-Catalyzed Substitution Reaction of Aryl Fluorides with Disulfides: *p*-Orientation in the Polyarylthiolation of Polyfluorobenzenes. *J. Am. Chem. Soc.* **2008**, *130*, 12214–12215. [CrossRef] [PubMed]
48. Arisawa, M.; Ono, T.; Yamaguchi, M. Rhodium-Catalyzed Thiophosphinylation and Phosphinylation Reactions of Disulfides and Diselenides. *Tetrahedron Lett.* **2005**, *46*, 5669–5671. [CrossRef]
49. Arisawa, M.; Igarashi, Y.; Kobayashi, H.; Yamada, T.; Bando, K.; Ichikawa, T.; Yamaguchi, M. Equilibrium Shift in the Rhodium-catalyzed Acyl Transfer Reactions. *Tetrahedron* **2011**, *67*, 7846–7859. [CrossRef]
50. Arisawa, M.; Fukumoto, K.; Yamaguchi, M. Rhodium-Catalyzed Phophorylation Reaction of Water-Soluble Disulfides Using Hypodiphosphoric Acid Tetraalkyl Esters in Water. *RSC Adv.* **2020**, *10*, 13820–13823. [CrossRef]

51. Arisawa, M.; Sawahata, K.; Yamada, T.; Sarkar, D.; Yamaguchi, Y. Rhodium-Catalyzed Insertion Reaction of PhP Group of Pentaphenylcyclopentaphosphine with Acyclic and Cyclic Disulfides. *Org. Lett.* **2018**, *20*, 938–941. [CrossRef]
52. Arisawa, M.; Tazawa, T.; Ichinose, W.; Kobayashi, H.; Yamaguchi, M. Rhodium-Catalyzed Synthesis of Dialkyl(Heteroaryl)Phosphine Sulfides by Phosphinylation of Heteroaryl Sulfides. *Adv. Synth. Catal.* **2018**, *360*, 3488–3491. [CrossRef]
53. Arisawa, M.; Watanabe, T.; Yamaguchi, M. Direct Transformation of Organosulfur Compounds to Organophosphorus Compounds: Rhodium-catalyzed Synthesis of 1-Alkynylphosphine Sulfides and Acylphosphine Sulfides. *Tetrahedron Lett.* **2011**, *52*, 2410–2412. [CrossRef]
54. Arisawa, M.; Yamaguchi, M. Addition Reaction of Dialkyl Disulfides to Terminal Alkynes Catalyzed by a Rhodium Complex and Trifluoromethanesulfonic Acid. *Org. Lett.* **2001**, *3*, 763–764. [CrossRef]
55. Arisawa, M.; Kozuki, Y.; Yamaguchi, M. Rhodium-Catalyzed Regio-and Stereoselective 1-Seleno−2-thiolation of 1-Alkynes. *J. Org. Chem.* **2003**, *68*, 8964–8967. [CrossRef]
56. Arisawa, M.; Suwa, A.; Fujimoto, K.; Yamaguchi, M. Transition Metal-Catalyzed Synthesis of (E)-2-(Alkylthio)-alka−1,3-dienes from Allenes and Dialkyl Disulfides with Concomitant Hydride Transfer. *Adv. Synth. Catal.* **2003**, *345*, 560–563. [CrossRef]
57. Zhang, H.; Wang, H.; Yang, H.; Fu, H. Rhodium-catalyzed Denitrogenative Thioacetalization of N-Sulfonyl−1,2,3-triazoles with Disulfides: An Entry to Diverse Transformation of Terminal Alkynes. *Org. Biomol. Chem* **2015**, *13*, 6149–6153. [CrossRef] [PubMed]
58. Khanal, H.D.; Kimb, S.H.; Lee, Y.R. Rhodium(II)-catalyzed Direct Sulfenylation of Diazooxindoles with Disulfides. *RSC Adv.* **2016**, *6*, 58501–58510. [CrossRef]
59. Arisawa, M.; Sugata, C.; Yamaguchi, M. Oxidation/Reduction Interconversion of Thiols and Disulfides Using Hydrogen and Oxygen Catalyzed by a Rhodium Complex. *Tetrahedron Lett.* **2005**, *46*, 6097–6099. [CrossRef]
60. Arisawa, M.; Tanaka, K.; Yamaguchi, M. Rhodium-Catalyzed Sulfur Atom Exchange Reaction between Organic Polysulfides and Sulfur. *Tetrahedron Lett.* **2005**, *46*, 4797–4800. [CrossRef]
61. Arisawa, M.; Ashikawa, M.; Suwa, A.; Yamaguchi, M. Rhodium-catalyzed Synthesis of Isothiocyanate from Isonitrile and Sulfur. *Tetrahedron Lett.* **2005**, *46*, 1727–1729. [CrossRef]
62. Adam, W.; Bargon, R.M.; Bosio, S.G.; Schenk, W.A.; Stalke, D. Direct Synthesis of Isothiocyanates from Isonitriles by Molybdenum-Catalyzed Sulfur Transfer with Elemental Sulfur. *J. Org. Chem.* **2002**, *67*, 7037–7041. [CrossRef]
63. Arisawa, M.; Ichikawa, T.; Yamaguchi, M. Rhodium-Catalyzed Synthesis of Diaryl Sulfides Using Aryl Fluorides and Sulfur/Organopolysulfides. *Org. Lett.* **2012**, *14*, 5318–5321. [CrossRef]
64. Adam, W.; Bargon, R.M.; Schenk, W.A. Direct Episulfidation of Alkenes and Allenes with Elemental Sulfur and Thiiranes as Sulfur Sources, Catalyzed by Molybdenum Oxo Complexes. *J. Am. Chem. Soc.* **2003**, *125*, 3871–3876. [CrossRef]
65. Arisawa, M.; Ichikawa, T.; Yamaguchi, M. Synthesis of Thiiranes by Rhodium-Catalyzed Sulfur Addition Reaction to Reactive Alkenes. *Chem. Commun.* **2015**, *51*, 8821–8824. [CrossRef]
66. Arisawa, M.; Ichikawa, T.; Tanii, S.; Yamaguchi, M. Synthesis of Symmetrical and Unsymmetrical 1,4-Dithiins by Rhodium-Catalyzed Sulfur Addition Reaction to Alkynes. *Synthesis* **2016**, *48*, 3107–3119. [CrossRef]
67. Arisawa, M.; Tagami, Y.; Yamaguchi, M. Two Types of Rhodium-Catalyzed CS/CS Metathesis Reactions: Formation of CS/CS Bonds and CC/SS Bonds. *Tetrahedron Lett.* **2008**, *49*, 1593–1597. [CrossRef]
68. Arisawa, M.; Igarashi, Y.; Tagami, Y.; Yamaguchi, M.; Kabuto, C. Rhodium-catalyzed Carbothiolation Reaction of 1-Alkylthio−1-Alkynes. *Tetrahedron Lett.* **2011**, *52*, 920–922. [CrossRef]
69. Li, G.; Arisawa, M.; Yamaguchi, M. Rhodium-Catalyzed Synthesis and Reactions of N-Acylphthalimides. *Asian J. Org. Chem.* **2013**, *2*, 983–988. [CrossRef]
70. Arisawa, M.; Nihei, Y.; Suzuki, T.; Yamaguchi, M. Rhodium-Catalyzed Cleavage Reaction of Aryl Methyl Ethers with Thioesters. *Org. Lett.* **2012**, *14*, 855–857. [CrossRef]
71. Arisawa, M.; Tazawa, T.; Tanii, S.; Horiuchi, K.; Yamaguchi, M. Rhodium-Catalyzed Synthesis of Unsymmetric Di(heteroaryl) Sulfides Using Heteroaryl Ethers and S-Heteroaryl Thioesters via Heteroarylthio Exchange. *J. Org. Chem* **2017**, *82*, 804–810. [CrossRef]
72. Arisawa, M.; Nakai, K.; Yamada, T.; Suzuki, R.; Yamaguchi, M. Synthesis of Cycloalkyl/Steroidal Heteroaryl Sulfides Using Rhodium-Catalyzed Heteroaryl Exchange Reaction. *Heterocycles* **2020**, *100*, 104–118. [CrossRef]

73. Arisawa, M.; Kuwajima, M.; Toriyama, F.; Li, G.; Yamaguchi, M. Rhodium-Catalyzed Acyl-Transfer Reaction between Benzyl Ketones and Thioesters: Synthesis of Unsymmetric Ketones by Ketone CO-C Bond Cleavage and Intermolecular Rearrangement. *Org. Lett.* **2012**, *14*, 3804–3807. [CrossRef]
74. Li, G.; Arisawa, M.; Yamaguchi, M. Rhodium-catalyzed Synthesis of Unsymmetrical Di (Aryl/Heteroaryl) methanes Using Aryl/Heteroarylmethyl Ketons via CO-C. Bond Cleavage. *Chem. Commun.* **2014**, *50*, 4328–4330. [CrossRef]
75. Arisawa, M.; Tanii, S.; Yamada, T.; Yamaguchi, M. Palladium-catalyzed Addition Reaction of Thioesters to Norbornenes. *Tetrahedron* **2015**, *71*, 6449–6458. [CrossRef]
76. Arisawa, M.; Li, G.; Yamaguchi, M. Rhodium-Catalyzed Synthesis of 2,3-Diaryl-1,4-Diketones via Oxidative Coupling of Benzyl Ketones Using α-Thioketone Oxidizing Reagent. *Tetrahedron Lett.* **2013**, *54*, 1298–1301. [CrossRef]
77. Arisawa, M.; Ichikawa, T.; Yamaguchi, M. Synthesis of Unsymmetrical Polyfluorinated Diaryl Dulfides by Rhodium-Catalyzed Aryl Exchange Reaction. *Tetrahedron Lett.* **2013**, *54*, 4327–4329. [CrossRef]
78. Arisawa, M.; Nihei, Y.; Yamaguchi, M. Rhodium-Catalyzed 2-Methylthiolation Reaction of Thiazoles/Oxazoles Using 2-(Methylthio) Thizole. *Heterocycles* **2015**, *90*, 939–949. [CrossRef]
79. Shi, G.; Khan, R.; Zhang, X.; Yang, Y.; Zhan, Y.; Li, J.; Luo, Y.; Fan, B. Rhodium-Catalyzed Direct *ortho* C–H Thiolation of Cyclic N-Sulfonyl Ketimines. *Asian J. Org. Chem.* **2020**, *9*, 788–792. [CrossRef]
80. Kang, Y.-S.; Zhang, P.; Li, M.-Y.; Chen, Y.-K.; Xu, H.-J.; Zhao, J.; Sun, W.-Y.; Yu, J.-Q.; Lu, Y. Ligand-Promoted Rh^{III}-Catalyzed Thiolation of Benzamides with a Broad Disulfide Scope. *Angew. Chem. Int. Ed.* **2019**, *58*, 9099–9103. [CrossRef] [PubMed]
81. Liu, C.; Fang, Y.; Wang, S.-Y.; Ji, S.-J. RhCl$_3$·3H$_2$O-Catalyzed Ligand-Enabled Highly Regioselective Thiolation of Acrylic Acids. *Acs. Catal.* **2019**, *9*, 8910–8915. [CrossRef]
82. Liu, C.; Fang, Y.; Wang, S.-Y.; Ji, S.-J. Highly Regioselective Rh^{III}-Catalyzed Thiolation of N-Tosyl Acrylamides: General Access to (Z)-β-Alkenyl Sulfides. *Org. Lett.* **2018**, *20*, 6112–6116. [CrossRef] [PubMed]
83. Yang, J.; Deng, B.; Guo, X.; Li, Z.; Xiang, H.; Zhou, X. Rhodium(III)-Catalyzed Thiolation of Azobenzenes. *Asian J. Org. Chem.* **2018**, *7*, 439–443. [CrossRef]
84. Zhu, F.; Wu, X.-F. Carbonylative Synthesis of 3-Substituted Thiochromenones via Rhodium-Catalyzed [3 + 2 + 1] Cyclization of Different Aromatic Sulfides, Alkynes, and Carbon Monoxide. *J. Org. Chem.* **2018**, *83*, 13612–13617. [CrossRef] [PubMed]
85. Nguyen, T.B. Recent Advances in Organic Reactions Involving Elemental Sulfur. *Adv. Synth. Cat.* **2017**, *359*, 1066–1130. [CrossRef]
86. Maity, S.; Karmakar, U.; Samanta, R. Regiocontrolled Direct C4 and C2-Methylthiolation of Indoles under Rhodium-catalyzed Mild Conditions. *Chem. Commun.* **2017**, *53*, 12197–12200. [CrossRef] [PubMed]
87. Xie, W.; Li, B.; Wang, B. Rh(III)-Catalyzed C7-Thiolation and Selenation of Indolines. *J. Org. Chem.* **2016**, *81*, 396–403. [CrossRef] [PubMed]
88. Wen, J.; Wu, A.; Wang, M.; Zhu, J. Rhodium(III)-Catalyzed Directed ortho-C–H Bond Functionalization of Aromatic Ketazines via C–S and C–C Coupling. *J. Org. Chem.* **2015**, *80*, 10457–10463. [CrossRef] [PubMed]
89. Yang, Y.; Hou, W.; Qin, L.; Du, J.; Feng, H.; Zhou, B.; Li, Y. Rhodium-Catalyzed Directed Sulfenylation of Arene C–H Bonds. *Chem. Eur. J.* **2014**, *20*, 416–420. [CrossRef] [PubMed]
90. Moon, S.; Kato, M.; Nishii, Y.; Miura, M. Synthesis of Benzo[*b*]thiophenes through Rhodium-Catalyzed Three-Component Reaction using Elemental Sulfur. *Adv. Synth. Catal.* **2020**, *362*, 1669–1673. [CrossRef]
91. Arisawa, M.; Sawahata, K.; Ichikawa, T.; Yamaguchi, M. Rhodium-Catalyzed Isomerization and Alkyne Exchange Reactions of 1,4-Dithiins via the 1,2-Ethenedithiolato Rhodium Complex. *Organometallics* **2018**, *37*, 3174–3180. [CrossRef]
92. Gal, A.W.; Gosselink, J.W.; Vollenbroek, F.A. The Oxidative Addition of Unsaturated Cyclic Five-membered Disulphides to RhCl(PPh$_3$)$_3$ and Pt(PPh$_3$)$_4$. *Inorg. Chim. Acta.* **1979**, *32*, 235–241. [CrossRef]
93. Seino, H.; Yoshikawa, T.; Hidai, M.; Mizobe, Y. Preparation of Mononuclear and Dinuclear Rh Hydrotris(pyrazolyl)borato Complexes Containing Arenethiolato Ligands and Conversion of the Mononuclear Complexes into Dinuclear Rh-Rh and Rh-Ir Complexes with Bridging Arenethiolato Ligands. *Dalton Trans.* **2004**, 3593–3600. [CrossRef]
94. De Croon, M.H.J.M.; van Gaal, H.L.M. Rhodium(I)- and Iridium-(1)-diethyldithiocarbamatoalkene Complexes and their Use in the Preparation of Rhodium(III)- and Iridium(III)-diethyldithiocarbamato Complexes. *Inorg. Nucl. Chem. Lett.* **1974**, *10*, 1081–1086. [CrossRef]

95. Ginsberg, A.P.; Lindsell, W.E.; Sprinkle, C.R.; West, K.W.; Cohen, R.L. Disulfur and Diselenium Complexes of Rhodium and Iridium. *Inorg. Chem.* **1982**, *21*, 3666–3681. [CrossRef]
96. Gangopadhyay, S.; Basak, P.; Drew, M.; Gangopadhyay, P.K. In Situ Formation of Ligand 2,2′-[(E)-diazene–1,2-diyldicarbonothioyl] diphenol and Structural Characterization of its Binuclear Rhodium(V) Complex Containing $RhO_2{}^+$. *Chem. Commun.* **2010**, *46*, 7436–7438. [CrossRef] [PubMed]
97. Duckett, S.B.; Haddleton, D.M.; Jackson, S.A.; Perutz, R.N.; Poliakoff, M.; Upmacis, R.K. Photochemical Oxidative Addition Reactions of (η^5-Cyclopentadienyl)-bis(ethene)rhodium with Dihydrogen and Trialkylsilanes: Formation and Isolation of Rhodlum(III) and Rhodium(V) Hydrides. *Organometallics* **1988**, *7*, 1526–1532. [CrossRef]
98. Ruiz, J.; Spencer, C.M.; Mann, B.E.; Taylor, B.F.; Maitlis, P.M. The Synthesis and Characterisation of Dihydridobis(trialkyltin)(pentamethylcyclopentadienyl)-rhodium(V) and -iridium(V) Complexes and Related Reactions. *J. Organomet. Chem.* **1987**, *325*, 253–260. [CrossRef]
99. Fernandez, M.-J.; Bailey, P.M.; Bentz, P.O.; Ricci, J.S.; Koetzle, T.F.; Maitlis, P.M. Synthesis, X-ray, and Low-Temperature Neutron Diffraction Study of a Rhodium(V) Complex: Dihydridobis(triethylsilyl)-pentamethylcyclopentadienylrhodium. *J. Am. Chem. Soc.* **1984**, *106*, 5458–5463. [CrossRef]
100. Tanaka, K.; Ajiki, K. Rhodium-Catalyzed Reaction of Thiols with Polychloroalkanes in the Presence of Triethylamine. *Org. Lett.* **2005**, *7*, 1537–1539. [CrossRef]
101. Japan Soda Industry Association. Available online: https://www.jsia.gr.jp/description/ (accessed on 5 August 2020).
102. The Federation of Electric Power Companies of Japan. Available online: https://www.fepc.or.jp/library/data/demand/__icsFiles/afieldfile/2016/04/28/juyou_k_fy2015.pdf (accessed on 5 August 2020).
103. Trost, B.M. The Atom Economy-A Search for Synthetic Efficiency. *Science* **1991**, *254*, 1471–1477. [CrossRef]

© 2020 by the authors. Licensee MDPI, Basel, Switzerland. This article is an open access article distributed under the terms and conditions of the Creative Commons Attribution (CC BY) license (http://creativecommons.org/licenses/by/4.0/).

Article
Reductive Hydroformylation of Isosorbide Diallyl Ether

Jérémy Ternel [1], Adrien Lopes [1,2], Mathieu Sauthier [2], Clothilde Buffe [3], Vincent Wiatz [3], Hervé Bricout [1], Sébastien Tilloy [1,*] and Eric Monflier [1]

1. University of Artois, CNRS, Centrale Lille, University of Lille, UMR 8181–UCCS–Unité de Catalyse et Chimie du Solide, F-62300 Lens, France; jeremy.ternel@univ-artois.fr (J.T.); adrien.figueiredo.lopes@gmail.com (A.L.); herve.bricout@univ-artois.fr (H.B.); eric.monflier@univ-artois.fr (E.M.)
2. University of Lille, CNRS, Centrale Lille, ENSCL, University of Artois, UMR 8181, Unité de Catalyse et Chimie du Solide, F-59000 Lille, France; mathieu.sauthier@univ-lille.fr
3. Roquette Frères, 1 Rue de la Haute Loge, F-62136 Lestrem, France; clothilde.buffe@roquette.com (C.B.); vincent.wiatz@roquette.com (V.W.)
* Correspondence: sebastien.tilloy@univ-artois.fr

Abstract: Isosorbide and its functionalized derivatives have numerous applications as bio-sourced building blocks. In this context, the synthesis of diols from isosorbide diallyl ether by hydrohydroxymethylation reaction is of extreme interest. This hydrohydroxymethylation, which consists of carbon-carbon double bonds converting into primary alcohol functions, can be obtained by a hydroformylation reaction followed by a hydrogenation reaction. In this study, reductive hydroformylation was achieved using isosorbide diallyl ether as a substrate in a rhodium/amine catalytic system. The highest yield in bis-primary alcohols obtained was equal to 79%.

Keywords: catalysis; hydroformylation; hydrogenation; rhodium; tandem reaction

1. Introduction

The depletion of fossil resources is a growing problem. Consequently, the challenge is to find renewable substitutable resources for the synthesis of chemical products. So, in the context of green chemistry and sustainable development, the synthesis of simple molecules from biomass with chemically transformable functions is an essential goal. Moreover, functionalization, such as the use of highly selective catalytic processes with a high economy of atoms, must be in line with the principles of green chemistry. The polymer industry represents a large part of the world fossil resource consumption. In this sense, the design of new monomers from bio-based molecules represents a major challenge concerning the reduction of fossil resources. Biomass is mainly composed of vegetable oils (5–13%), lignin (15–25%), and carbohydrates (61–82%) [1]. The valorization of this last segment, mainly present in biomass, makes it possible to obtain polyfunctional molecules such as furans, lactic acid, and, more particularly, sorbitol obtained from glucose, which can be accessible from starch. Sorbitol and its isomers can be dehydrated to form dianhydrohexitols that bear bicyclic scaffolds and secondary diols. Among the possible isomers, isosorbide is produced on an industrial scale, with the Roquette company being one of the largest producers. The literature describes numerous bio-sourced compounds accessible from isosorbide, which have multiple applications, such as in the medical field for isosorbide dinitrates [2,3], in non-toxic green solvents for dimethyl isosorbide [4], and, more particularly, in polymer chemistry. Many reviews list a large number of biopolymers derived from isosorbide or its functionalized derivatives [5–8]. Isosorbide is mainly employed as a monomer in the production of polyesters and polycarbonates [9–12]. Isosorbide derivatives, where the hydroxyl groups are replaced by amine, carboxylic acid, or nitrile functions are described [13–15]. Other derivatives, where the reactive function or a carbon-carbon double bond are carried by an extra-cyclic carbon, have also been synthesized. More particularly,

the synthesis of isosorbide diallyl ether has been extensively studied for its transformation into isosorbide bis-glycidyl ether (Bisglycidyl Ether) by epoxidation of the double bonds [16]. Although the isosorbide bis-glycidyl ether is often described in the literature, little attention has been paid to other possible transformations of the isosorbide diallyl ether derivative by functionalization of the carbon-carbon double bonds. More precisely, organometallic catalysis offers many opportunities for incorporating polymerisable functions such as carboxylic acid, ester, aldehyde, amine, or alcohol on allyl derivatives of isosorbide. In this context, we were interested in the synthesis of diols from isosorbide diallyl ether by hydrohydroxymethylation (HHM) reaction. HHM consists of the conversion of carbon-carbon double bonds into primary alcohol functions via a hydroformylation reaction, followed by a hydrogenation reaction [17]. These primary alcohol functions can be produced in one- or two-step processes by using one or two catalysts, and under similar or different reaction conditions [18–21]. The process in a one-step reaction with a single catalyst is the best alternative, owing to the reduction in the amount of the catalyst, the reaction time, and the overall cost. In this context, one of the best systems is a Rh-catalyst associated with tertiary amines in the presence of CO/H_2 [22–33]. In this publication, we report the HHM of isosorbide diallyl ether using this catalytic system. The effects of various reaction parameters (nature and concentration of amines, time reaction, and syngas composition) were studied in order to increase the production of primary alcohols.

2. Results

2.1. Presentation of the Compounds Resulting from HHM of IDE

The isosorbide diallyl ether (IDE) used in the study was synthesized by the reaction of isosorbide with allyl bromide in a basic aqueous medium in a similar fashion to previous publications [34,35]. The hydrohydroxymethylation of isosorbide diallyl ether (IDE) by the [Rh(acac)(CO)$_2$/amine] catalytic system under CO/H_2 pressure can produce many compounds. Scheme 1 summarized the theoretical compounds. In particular, isosorbide diallylic ether was expected to be hydrohydroxymethylated into bis-primary alcohols. However, due to the nature of the allylated isosorbide substrate and the isomerizing, hydroformylating, and hydrogenating properties of the [Rh(acac)(CO)$_2$/amine] catalytic system under CO/H_2 pressure, many products could be obtained under catalytic conditions (Scheme 1a). More precisely, each allyl group (2-propenyl group (**2-P**)) of the allylated isosorbide could be converted into a propyl group (**P**) by hydrogenation of the carbon-carbon double bond, or isomerized into a *(Z)*- or an *(E)*-1-propenyl group (**1-P**) (Scheme 1b).

Because this (**1-P**) group proved to be completely unreactive in our experimental conditions (see Section 2.2 and Scheme 2), its transformation into (**P**) group by hydrogenation and to 1-formylpropyl (**1-FP**) and 2-formylpropyl (**2-FP**) groups by hydroformylation was excluded. In the same way, because the (**1-FP**) group was not formed, its 1-(hydroxymethyl)propyl (**1-HMP**) corresponding hydrogenated form was also not taken into account.

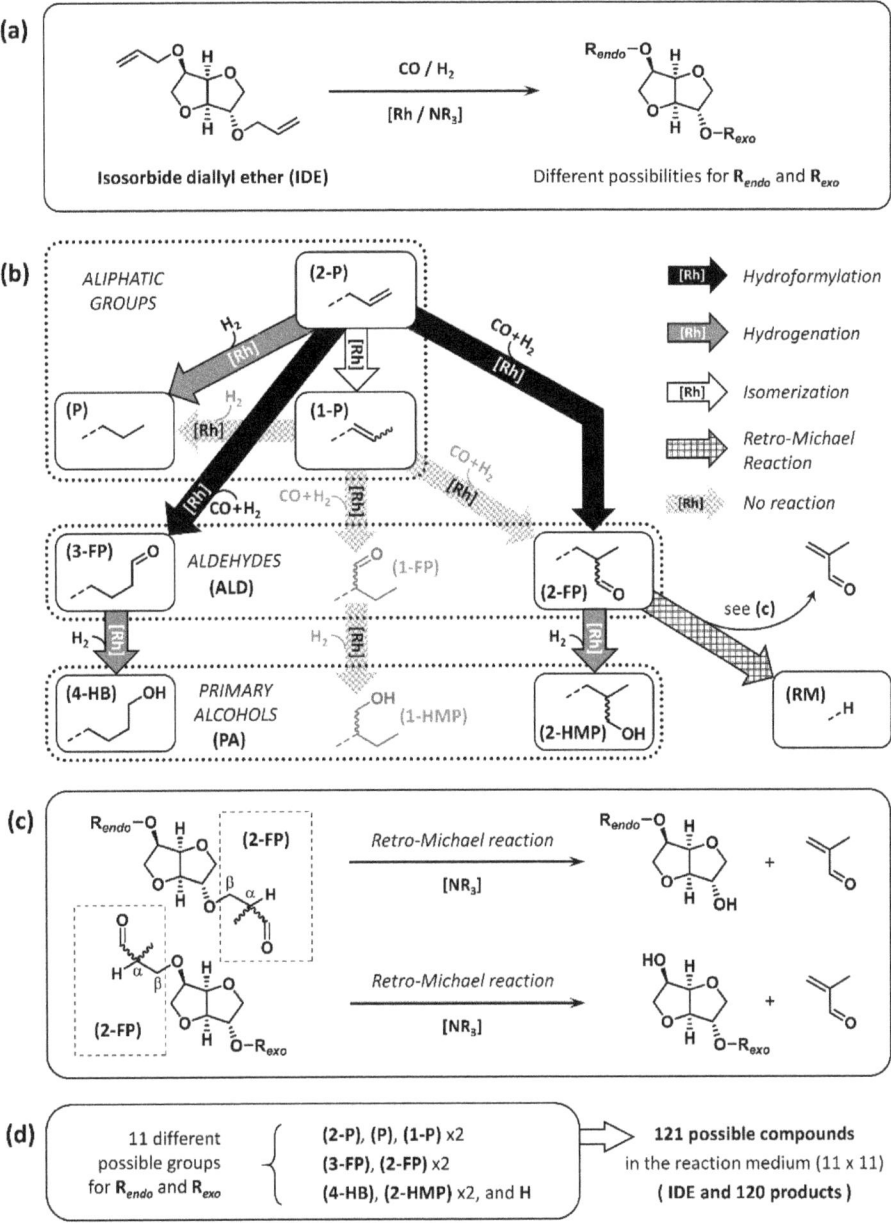

Scheme 1. Reactions of isosorbide diallyl ether (IDE) in the presence of the [Rh(acac)(CO)$_2$/NR$_3$] catalytic system and under CO/H$_2$ pressure. (**a**) General equation; (**b**) different possible groups linked to the isosorbide moiety; (**c**) two possible retro-Michael reactions; and (**d**) determination of the total number of possible compounds in the reaction medium. The various groups are allyl (**2-P**), propyl (**P**), *(Z)*- or *(E)*-1-propenyl (**1-P**), 3-formylpropyl (**3-FP**), *(R)*- or *(S)*-2-formylpropyl (**2-FP**), 4-hydroxybutyl (**4-HB**), *(R)*- or *(S)*-2-(hydroxymethyl)propyl (**2-HMP**), and a hydrogen atom (**RM**). (**1-FP**) and (**1-HMP**) groups are not formed because (**1-P**) group was proven to be unreactive in our experimental conditions (see Scheme 2).

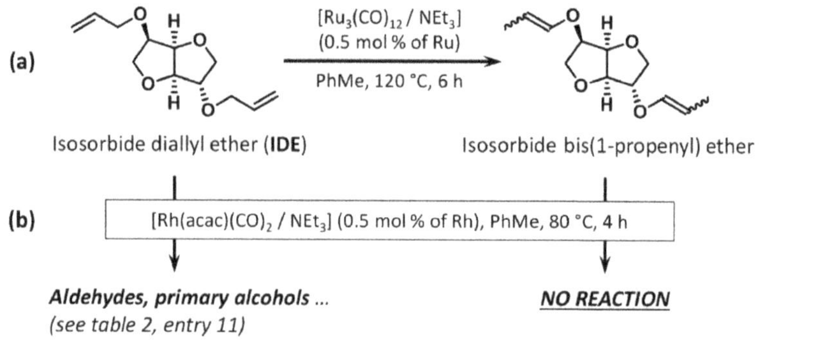

Scheme 2. (a) Synthesis of isosorbide bis(1-propenyl) ether from isosorbide diallyl ether and (b) comparison of these two compounds behaviors under rhodium-catalyzed hydrohydroxymethylation conditions.

On the other hand, each allyl (**2-P**) group could be hydroformylated into a linear or branched aldehyde graft, namely, 3-formylpropyl (**3-FP**) and 2-formylpropyl (**2-FP**) groups. The two aldehyde grafts (**3-FP**) and (**2-FP**) ((**ALD**) for the general term) could then be hydrogenated into 4-hydroxybutyl (**4-HB**) and 2-(hydroxymethyl)propyl (**2-HMP**) primary alcohol groups, respectively ((**PA**) for the general term). Furthermore, the aldehyde function of the (**2-FP**) graft showed a hydrogen atom and an oxygen atom on the α and β positions, respectively (Scheme 1c). So, in the presence of the basic amine ligand, a retro-Michael reaction could not be excluded [36]. This reaction would lead to the cleavage of the (**2-FP**) graft with the formation of 2-methylacrolein and the initial ether-oxide function would be transformed into a secondary alcohol function on the isosorbide moiety (Scheme 1b,c).

In other words, the (**2-FP**) graft was replaced by a hydrogen atom (named (**RM**) for retro-Michael). Because two stereochemical forms were possible for the (**1-P**) graft (Z and E), as for (**2-FP**) and (**2-HMP**) grafts (R and S), 11 different substituents, bearing by the isosorbide moiety, were possible (Scheme 1d): 4 aliphatic groups ((**2-P**), (**P**), (Z)-(**1-P**) and (E)-(**1-P**)), 3 aldehyde groups ((**ALD**) = (**3-FP**), (R)-(**2-FP**), (S)-(**2-FP**)), 3 primary alcohol groups ((**PA**) = (**4-HB**), (R)-(**2-HMP**), (S)-(**2-HMP**)), and the hydrogen atom coming from the retro-Michael reaction ((**RM**)). Because of these 11 possibilities for R_{endo} and R_{exo} groups on the isosorbide moiety, 121 compounds could be present in the reaction medium (11 × 11), including the IDE substrate and 120 reaction products (Scheme 1d).

2.2. Preliminary Reactions

In order to illustrate the proof of the concept presented earlier, some exploratory experiments were performed. The applied experimental conditions were inspirited by our previous works for HHM of triglycerides [31]. More precisely, the reaction was carried out by using Rh(acac)(CO)$_2$ (0.5 mol%) as a rhodium precursor, triethylamine (TEA, 200 eq.) as ligand, and toluene (6 mL) as a solvent under 80 bars of CO/H$_2$ at 80 °C during 4 h. For these preliminary reactions, two substrates were envisaged: isosorbide diallyl ether (IDE, isosorbide bearing (**2-P**) groups) and isosorbide bis(1-propenyl) ether (isosorbide bearing (**1-P**) groups). This last substrate was prepared from IDE by using a Ru/TEA catalytic system to isomerize the carbon-carbon double bonds (see Scheme 2; see also part III.2 of Supplementary Materials for exact experimental conditions).

As previously explained, no conversion of the 1-propenyl groups ((**1-P**)) of the isosorbide bis(1-propenyl) ether was observed under HHM reaction conditions. This result was very interesting because it proved that the (**1-FP**) and (**1-HMP**) grafts, potentially formed from the allyl group after isomerization into (**1-P**) (see Scheme 1), would not appear in the reactions involving IDE as substrate.

In the case of IDE as substrate, the application of the same HHM reaction conditions used for unreactive isosorbide bis(1-propenyl) ether led to numerous products. Figure 1 presents the ^{1}H NMR spectrum of the mixture of products obtained after 4 h of reaction (this experiment corresponds to entry 11 of Table 1). The different characteristic ^{1}H NMR signals were identified by comparison with authentic samples (see part III in Supporting Materials). The chemical shift zones that have been used for allyl group conversion and yield determination are described in Supporting Materials (see part IV.2). More precisely, in order to facilitate the description of the results, three items were determined for each catalytic test: (i) the allyl group conversion; (ii) the corresponding yields in the different hydrogenated, isomerized, hydroformylated, and hydrohydroxymethylated allyl groups; and (iii) the yield in bis-primary alcohols, with respect to the initial isosorbide moiety. The allyl group conversion and corresponding yields were determined from the ^{1}H NMR spectrum, while the yield in bis-primary alcohols was determined by gas chromatography (see parts IV.2 and IV.3 in Supplementary Materials).

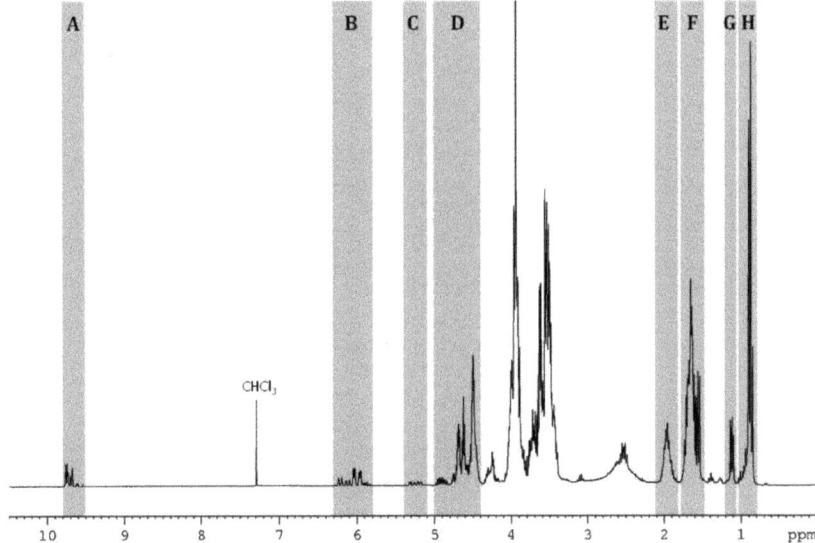

Figure 1. ^{1}H NMR spectrum of the final HHM reaction medium of isosorbide diallyl ether (IDE) after evaporation of triethylamine and toluene (300 MHz, CDCl$_3$, 25 °C); experimental conditions: Rh(acac)(CO)$_2$ (12.9 mg, 50 µmol, 1 equiv), TEA (200 equiv), IDE (2.26 g, 10 mmol, 200 equiv), toluene (6 mL), 80 bars of CO/H$_2$ (1:1), 80 °C, 4 h; the values of integration of the different chemical shift zones (zones **A** to **H**) were used to determine conversion and yields (see part IV.2 of Supplementary Materials).

Table 1. Influence of the TEA amount on IDE hydrohydroxymethylation [a].

Entry	TEA/Rh	t (h)	Conv. [b] (%)	$Y_{(P)}$ [c] (%)	$Y_{(1-P)}$ [c] (%)	$Y_{(ALD)}$ [c] (%) [l/b] [d]	$Y_{(PA)}$ [c] (%) [l/b] [e]	$Y_{(RM)}$ [c] (%)	$Y'_{(BPA)}$ [f] (%) [ll/lb/bb] [g]
1 [h]	0	18	5	0	5	0	0	0	0
2	0	18	100	9	3	82 [50/50]	0	6	0
3	10	6	95	2	12	21 [45/55]	54 [38/62]	6	38 [12/40/48]
4	20	6	100	2	4	22 [45/55]	67 [40/60]	5	44 [12/46/42]
5	50	6	100	2	7	6 [49/51]	80 [45/55]	5	63 [17/49/34]
6	100	6	100	2	5	0	87 [45/55]	6	67 [17/50/33]
7	200	6	100	2	5	0	87 [44/56]	6	69 [17/49/34]
8 [i]	1200	18	100	3	6	0	85 [43/57]	6	62 [16/49/35]
9	20	4	79	2	3	38 [50/50]	30 [40/60]	5	22 [12/43/45]
10	20	18	100	3	4	0	89 [46/54]	4	67 [17/50/33]
11	200	4	95	1	4	16 [48/52]	70 [46/54]	4	49 [17/48/35]
12	200	18	100	3	5	0	86 [45/55]	6	67 [17/79/34]

[a] Experimental conditions: Rh(acac)(CO)$_2$ (12.9 mg, 50 μmol, 1 equiv), IDE (2.26 g, 10 mmol, 200 equiv), TEA (0–1200 equiv), toluene (6 mL), 80 bars of CO/H$_2$ (1:1), 80 °C. [b] Conv. = allyl groups conversion determined by ^{1}H NMR. [c] $Y_{(X)}$ = yield in (X) with respect to the initial allyl group, determined by ^{1}H NMR; **(P)** = propyl; **(1-P)** = 1-propenyl; **(ALD)** = aldehyde groups = **(2-FP)** + **(3-FP)**; **(PA)** = primary alcohol groups = **(2-HMP)** + **(4-HB)**; **(RM)** = **(2-FP)** grafts that have been cleaved by retro-Michael reaction. [d] Linear to branched ratio for aldehydes i.e., **(3-FP)**/**(2-FP)** molar ratio. [e] Linear to branched ratio for primary alcohols i.e., **(4-HB)**/**(2-HMP)** molar ratio. [f] $Y'_{(BPA)}$ = yield in bis-primary alcohols with respect to the initial isosorbide moiety, determined by gas chromatography (GC-FID). [g] linear-linear/linear-branched/branched-branched ratios for bis-primary alcohols, determined by gas chromatography (GC-FID). [h] 1 bar of N$_2$ instead of 80 bars of CO/H$_2$. [i] 30 mmol of IDE instead of 10, and no toluene.

In this way, the analysis of the ^{1}H NMR spectrum of Figure 1 showed that the preliminary HHM experiment performed with IDE as a substrate gave the following results after 4 h of reaction. The allyl group conversion was equal to 95%. The different related yields were equal to 1% for **(P)** group ($Y_{(P)}$(%) = 1), 4% for **(1-P)** group ($Y_{(1-P)}$(%) = 4), 16% for aldehyde groups ($Y_{(ALD)}$(%) = 16), 4% for hydrogen group (retro-Michael reaction; $Y_{(RM)}$(%) = 4), and 70% of primary alcohol groups ($Y_{(PA)}$(%) = 70; with a distribution **(4-HB)**/**(2-HMP)** = 46/54). Furthermore, the analysis of the final reaction medium by gas chromatography indicated a yield in bis-primary alcohols equal to 49%, with respect to the initial isosorbide moiety ($Y'_{(BPA)}$(%) = 49).

2.3. HHM of IDE: Effect of the TEA Amount and Reaction Time

By following the described experimental conditions, the effect of TEA amount and reaction time was studied (Table 1). An experiment performed without TEA and CO/H$_2$ resulted in only isomerization products with a poor allyl group conversion (5%) in 18 h (Table 1, entry 1). The same experiment conducted under CO/H$_2$ provided hydrogenated, isomerized, hydroformylated, and retro-Michael products, but no alcohol, after 18 h (Table 1, entry 2).

By increasing the TEA amount from 10 to 200 equivalents with respect to Rh, allyl group conversions quickly reached 100% in 6 h (from a ratio of 20; Table 1, entries 3–7). From a ratio TEA/Rh of 100, the complete conversion of aldehydes to alcohols was observed. Interestingly, the yields in primary alcohols ($Y_{(PA)}$) reached 87% (with a linear to branched alcohol ratio around l/b = 44/56; l for **(4-HB)** group, and b for **(2-HMP)** group). Logically, good yields in bis-primary alcohols were also obtained ($Y'_{(BPA)}$ = 67–69%). The regioselectivity related to the primary alcohol group inside of the bis-primary alcohol products was around 17/49/34 for linear-linear(ll)/linear-branched(lb)/branched-branched(bb) ratio, respectively. The products issued from the retro-Michael reaction were always lower than 6%, showing this reaction was minor. Without toluene and with 30 mmol of IDE instead of 10, a total conversion and a yield of 62% in bis-primary alcohols were reached after 18 h (Table 1, entry 8).

The effect of the reaction time was studied at ratio TEA/Rh of 20 and 200 for 4 h and 18 h, to be compared to 6 h (Table 1; entries 9 and 10 to be compared to entry 4 for TEA/Rh = 20, and entries 11 and 12 to be compared to entry 7 for TEA/Rh = 200). For both TEA/Rh ratios, a shorter reaction time (4 h) was not sufficient to reach a total allyl group conversion (see entries 9 and 11). For the TEA/Rh ratio of 20, an increase in reaction time from 6 h to 18 h allowed for a transformation of the residual aldehydes groups present at 6 h ($Y_{(ALD)}$(%) = 22, entry 4) into primary alcohol groups, reaching a primary alcohol yield of 89% ($Y_{(PA)}$(%) = 89, entry 10; see also the kinetic follow-up of this reaction in Supplementary Materials, part IV.4, Figure S29). For the TEA/Rh ratio of 200, the medium composition stayed logically unchanged between 6 h and 18 h because, at 6 h, all the aldehydes were already completely transformed and the CC double bonds still present at t = 6 h were not reactive, (belonging to **(1-P)** groups; compare entries 7 and 12).

2.4. HHM of IDE: Effect of the Nitrogen Ligand

Instead of TEA, other trialkylamines (monodentate TBA and bidentates TMEDA, TMPDA, and TMBDA) and an aromatic amine (TPA) were used as Rh-ligands in the HHM of IDE (Scheme 3). The catalytic result obtained in the presence of TEA as ligand and after 6 h of reaction time was evoked for the comparison (Table 2, entry 1). In the presence of TPA (pKa = −3), no primary alcohol was produced (Table 2, entry 2). This behavior was probably due to the low pKa value. Indeed, some works relative of the HHM of olefins showed that the rhodium/amine complexes were only efficient to produce alcohols for pKa values comprised between 7 and 11 [29]. The other tested amines met this requirement. Consequently, primary alcohols were well produced by using TEA ($Y_{(PA)}$ = 87%; Table 2 entry 1), TBA ($Y_{(PA)}$ = 84%; Table 2, entry 3), TMEDA ($Y_{(PA)}$ = 90%; Table 2, entry 4), TMPDA ($Y_{(PA)}$ = 89%; Table 2, entry 5), and TMBDA ($Y_{(PA)}$ = 91%; Table 2, entry 6) at a constant nitrogen/Rh molar ratio of 200 (diamines/Rh ratios were equal to 100).

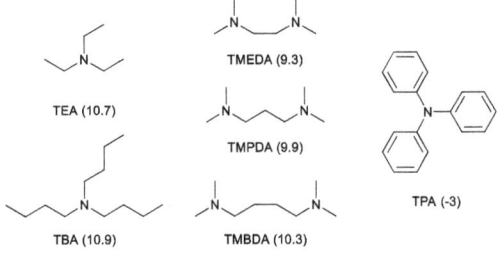

TEA	Triethylamine	TMEDA	N,N,N',N'-tetramethyl-1,2-ethanediamine
TBA	Tributylamine	TMPDA	N,N,N',N'-tetramethyl-1,3-propanediamine
TPA	Triphenylamine	TMBDA	N,N,N',N'-tetramethyl-1,4-butanediamine

Scheme 3. Amines and diamines used (pKa in brackets).

For these tertiary alkyl-amines and -diamines active in aldehydes hydrogenation, no significant change was observed for the yields in isomerized, saturated, and hydrogen groups (retro-Michael); the yields stayed lower than 10% (entries 1 and 3–6). It is also interesting to notice that the distribution of (*ll*)-, (*lb*)-, and (*bb*)-bis-primary alcohols was not modified in the presence of the bidentate diamines. This behavior was unexpected. Indeed, a modification of the distribution of linear/branched products versus the geometry of the ligand could have been observed. Indeed, in the case of phosphine as a ligand, this phenomenon is widely described in the literature [37], but compared to phosphines, amines are less efficient ligands.

Table 2. Influence of the nitrogen compound nature on IDE hydrohydroxymethylation [a].

Entry	Ligand (equiv/Rh)	Conv. [b] (%)	$Y_{(P)}$ [c] (%)	$Y_{(1-P)}$ [c] (%)	$Y_{(ALD)}$ [c] (%) [l/b] [d]	$Y_{(PA)}$ [c] (%) [l/b] [e]	$Y_{(RM)}$ [c] (%)	$Y'_{(BPA)}$ [f] (%) [ll/lb/bb] [g]
1	TEA (200)	100	2	5	0	87 [44/56]	6	69 [17/49/34]
2	TPA (200)	100	9	2	80 [50/50]	0	9	0
3	TBA (200)	100	5	5	0	84 [42/58]	6	60 [16/48/36]
4	TMEDA (100)	100	1	3	0	90 [46/54]	6	74 [16/49/35]
5	TMPDA (100)	100	2	4	0	89 [47/53]	5	75 [15/48/37]
6	TMBDA (100)	100	1	1	0	91 [46/54]	7	79 [16/49/37]
7	TMEDA (200)	100	2	3	0	90 [46/54]	5	75 [15/48/37]
8	TMPDA (200)	100	1	3	0	90 [47/53]	6	76 [16/48/36]
9	TMBDA (200)	100	2	2	0	91 [46/54]	5	79 [16/49/37]

[a] Experimental conditions: Rh(acac)(CO)$_2$ (12.9 mg, 50 μmol, 1 equiv), IDE (2.26 g, 10 mmol, 200 equiv), Nitrogen compound (10–200 equiv), toluene (6 mL), 80 bars of CO/H$_2$ (1:1), 80 °C, 6 h. [b] Conv. = allyl groups conversion determined by ^{1}H NMR. [c] $Y_{(X)}$ = yield in (X) with respect to the initial allyl group, determined by ^{1}H NMR; (P) = propyl; (1-P) = 1-propenyl; (ALD) = aldehyde groups = (2-FP) + (3-FP); (PA) = primary alcohol groups = (2-HMP) + (4-HB); (RM) = (2-FP) grafts that have been cleaved by the retro-Michael reaction. [d] Linear to branched ratio for aldehydes i.e., (3-FP)/(2-FP) molar ratio. [e] Linear to branched ratio for primary alcohols i.e., (4-HB)/(2-HMP) molar ratio. [f] $Y'_{(BPA)}$ = yield in bis-primary alcohols with respect to the initial isosorbide moiety, determined by gas chromatography (GC-FID). [g] Linear-linear/linear-branched/branched-branched ratios for bis-primary alcohols, determined by gas chromatography (GC-FID).

For the diamines, the experiments with a ratio ligand/Rh equal to 200 (nitrogen/Rh of 400) were also performed, but the results were very similar (compare entries 4–6 and 7–9). Nevertheless, the best results in bis-primary alcohol production were obtained with the bidentate diamines ($Y'_{(BPA)}$ = 74–79%, to be compared to $Y'_{(BPA)}$ = 69% for TEA).

2.5. HHM of IDE: Effect of the Reaction Pressure

As the reaction sequence from alkene to alcohol consumed CO and H$_2$ with a ratio of 1:2, experiments with this ratio were performed. In the presence of TEA, similar results were obtained for CO/H$_2$ ratios of 1:1 or 1:2 (Table 3, entries 1 and 2). By decreasing the pressure from 80 to 40 bars, the yields in primary alcohol decreased from 87 to 48 (Table 3, entries 2–4). So, a pressure of 80 bars seemed appropriate. The influence of the CO:H$_2$ molar ratio was also studied for TMPDA and TMBDA diamines (Table 3, entries 5–8). As for TEA, no modification of the bis-primary alcohol yields was evidenced. In the same way, the variation of the pressure did not impact the distribution of (*ll*)-, (*lb*)-, and (*bb*)-bis-primary alcohols.

Table 3. Influence of syngas pressure and composition on IDE hydrohydroxymethylation [a].

Entry	Ligand (equiv/Rh)	P (bar) (CO:H_2)	Conv.[b] (%)	$Y_{(P)}$[c] (%)	$Y_{(1-P)}$[c] (%)	$Y_{(ALD)}$[c] (%) [l/b][d]	$Y_{(PA)}$[c] (%) [l/b][e]	$Y_{(RM)}$[c] (%)	$Y'_{(BPA)}$[f] (%) [ll/lb/bb][g]
1	TEA (200)	80 (1:1)	100	2	5	0	87 [44/56]	6	69 [17/49/34]
2	TEA (200)	80 (1:2)	100	3	4	0	87 [45/55]	6	69 [17/48/35]
3	TEA (200)	60 (1:1)	100	2	13	0	80 [48/52]	5	60 [20/51/29]
4	TEA (200)	40 (1:1)	85	2	18	13 [48/52]	48 [45/55]	4	22 [17/48/35]
5	TMPDA (100)	80 (1:1)	100	2	4	0	89 [47/53]	5	75 [15/48/37]
6	TMPDA (100)	80 (1:2)	100	3	2	0	88 [48/52]	7	74 [16/49/37]
7	TMBDA (100)	80 (1:1)	100	1	1	0	91 [46/54]	7	79 [16/49/37]
8	TMBDA (100)	80 (1:2)	100	3	3	0	87 [46/54]	7	78 [15/49/36]

[a] Experimental conditions: Rh(acac)(CO)$_2$ (12.9 mg, 50 µmol, 1 equiv), IDE (2.26 g, 10 mmol, 200 equiv), Nitrogen compound (100–200 equiv), toluene (6 mL), CO/H$_2$ (1:1 or 1:2; 40–80 bar), 80 °C, 6 h. [b] Conv. = allyl groups conversion determined by ^{1}H NMR. [c] $Y_{(X)}$ = yield in (X) with respect to the initial allyl group, determined by ^{1}H NMR; (P) = propyl; (1-P) = 1-propenyl; (ALD) = aldehyde groups = (2-FP) + (3-FP); (PA) = primary alcohol groups = (2-HMP) + (4-HB); (RM) = (2-FP) grafts that have been cleaved by the retro-Michael reaction. [d] Linear to branched ratio for aldehydes i.e., (3-FP)/(2-FP) molar ratio. [e] Linear to branched ratio for primary alcohols i.e., (4-HB)/(2-HMP) molar ratio. [f] $Y'_{(BPA)}$ = yield in bis-primary alcohols with respect to the initial isosorbide moiety, determined by gas chromatography (GC-FID). [g] Linear-linear/linear-branched/branched-branched ratios for bis-primary alcohols, determined by gas chromatography (GC-FID).

3. Materials and Methods

All reactions involving metal-amine catalysts were performed under an air atmosphere. The catalytic precursor Rh(acac)(CO)$_2$ was purchased from Strem Chemicals and used as received. The different amines were purchased from Acros or Aldrich and used without prior purification. Isosorbide was supplied by Roquette. Syngas (CO:H$_2$/1:1) and dihydrogen were provided by the Linde Group in cylinders pressurized at 200 bars. The catalytic experiments were conducted under a fume hood in a room equipped with a CO detector and an explosimeter, both connected to an alarm.

^{1}H and ^{13}C NMR spectra were recorded at 298 K on a Bruker Avance III HD 300(Wissembourg, France) NanoBay spectrometer equipped with a 5 mm broadband probe BBFO with Z-gradients, operating at 7.05 T field strength (300 MHz for ^{1}H nuclei and 75 MHz for ^{13}C nuclei). ^{1}H and ^{13}C chemicals shifts were determined using residual signals of the deuterated solvents and were calibrated vs. SiMe$_4$. Assignment of the signals was carried out using 1D (^{1}H, ^{13}C) and 2D (COSY, HMBC, HMQC) NMR experiments.

Gas chromatography with flame ionization detection (GC-FID) was monitored by analyzing aliquots of the reaction mixture using a Shimadzu GC-2010 Plus apparatus (Noisiel, France) equipped with an RTX-5 capillary column (30 m, 0.25 mm, 0.25 µm). The oven temperature was programmed as follows: initial temperature of 50 °C, increased to 250 °C by 15 °C/min and held for 15 min. The injector and detector temperatures were 250 °C and nitrogen was used as carrier gas at a constant column flow rate of 1.50 mL/min. An aliquot of the sample was injected in split mode.

Products were also analyzed by ESI-MS (electrospray ionization-mass spectrometry) using an AB SCIEX TripleTOF® 5600 mass spectrometer (AB Sciex, Singapore).

Fourier transform infrared spectroscopy (FT-IR) experiments were carried out in the 4000–400 cm^{-1} region with a spectral resolution of 2 cm^{-1} using a Shimadzu IR Prestige-21 spectrometer equipped with a PIKE MIRacle diamond crystal (Noisiel, France).

In a typical catalytic experiment, Rh(acac)(CO)$_2$ (12.9 mg, 0.05 mmol, 1 eq.), Triethylamine (1.012 g, 10 mmol, 200 eq.), isosorbide diallyl ether (2.261 g, 10 mmol, 200 eq.), and toluene (6 mL) were added in a 25 mL stainless-steel autoclave (Parr instrument company) equipped with a mechanical stirrer. The reactor was sealed, the reaction mixture was stirred, and the reactor was heated at 80 °C. Then, the reactor was pressurized with 80 bars of CO/H$_2$ (1:1). After the appropriate reaction time, the reactor was cooled to room temperature and depressurized. The crude mixture was concentrated to remove the TEA and the solvent. The mixture was then analyzed by ^{1}H NMR spectroscopy and GC-FID. All runs were performed at least twice in order to ensure reproducibility.

4. Conclusions

The direct functionalization of isosorbide diallyl ether with primary alcohol functions was performed efficiently by the [Rh(acac)(CO)$_2$/trialkylamine] catalytic system under CO/H$_2$ pressure. The highest yields in bis-primary alcohols were obtained with bidentate diamines, with a maximum value of 79% obtained with TMBDA. The nature of the ligand (monodentate or bidentate), as well as the ligand/Rh and CO/H$_2$ ratios, did not modify the linear and branched products distribution. Among these observations, the more surprising result was the absence of different behaviors between monodentate or bidentate amines concerning regioselectivity. Indeed, for all active amines tested, the branched 2-(hydroxymethyl)propyl group systematically formed in a majority compared to the linear 2-hydroxybutyl group with a ratio l/b of about 45/55. This ratio was rather consistent with a low hindered catalytic species. These results raised questions on the role of amines as a ligand in this reaction, the key factor relative to the amine efficiency being also linked to its pKa value. In this context, some studies are being run to clarify the exact role played by the amine in the catalytic system. Finally, as the introduction of primary alcohol functions allows circumvention of the low overall reactivity of the secondary alcohol groups of isosorbide after a preliminary allylation, the new isosorbide derivatives obtained by this method could be advantageously used to synthesize isosorbide bio-based polyesters and polyurethanes.

Supplementary Materials: The following are available online. The Supplementary Materials contain: a general purpose (part I); an experimental section with the isosorbide diallyl ether synthesis and the general procedure for the reductive hydroformylation catalytic tests (part II); a part dedicated to synthesis and characterization of authentic samples/reaction products, including hydrogenated, isomerized, hydroformylated, and hydrohydroxymethylated isosorbide diallyl ether (part III); and a part explaining the way by which the allyl group conversion, related yields, and regioselectivities were determined by ^{1}H NMR analysis; as well as the bis-primary alcohols yield and characteristics by gas chromatography (part IV).

Author Contributions: Conceptualization, M.S., V.W., S.T., and E.M.; methodology, J.T., A.L., C.B., and H.B.; validation, M.S., V.W., H.B., S.T., and E.M.; formal analysis, J.T., A.L., C.B., and H.B.; investigation, J.T., A.L., and H.B.; resources, M.S., C.B., V.W., H.B., S.T., and E.M.; data curation, J.T., A.L., and H.B.; writing—original draft preparation, M.S, H.B., S.T., and E.M.; writing—review and editing, M.S., V.W., H.B., S.T., and E.M.; supervision, M.S., V.W., S.T., and E.M.; project administration, M.S., V.W., and S.T.; funding acquisition, M.S., and S.T. All authors have read and agreed to the published version of the manuscript.

Funding: This research was funded by Agence Nationale de la Recherche (ANR) under the "Investissements d'Avenir" Program, grant number: ANR-10-IEED-0004-01.

Institutional Review Board Statement: Not applicable.

Informed Consent Statement: Not applicable.

Data Availability Statement: The data presented in this study are available in Supplementary Material.

Acknowledgments: Nicolas Kania, Johan Hachani, and Dominique Prévost are acknowledged for GC and IR characterizations, ESI-MS analyses, and technical assistance, respectively.

Conflicts of Interest: The authors declare no conflict of interest.

Sample Availability: Samples of the compounds are available from the authors.

References

1. Bozell, J.J.; Petersen, G.R. Technology development for the production of biobased products from biorefinery carbohydrates—the US Department of Energy's "Top 10" revisited. *Green Chem.* **2010**, *12*, 539–554. [CrossRef]
2. Cole, R.T.; Kalogeropoulos, A.P.; Georgiopoulou, V.V.; Gheorghiade, M.; Quyyumi, A.; Yancy, C.; Butler, J. Hydralazine and Isosorbide Dinitrate in Heart Failure Historical Perspective, Mechanisms, and Future Directions. *Circulation* **2011**, *123*, 2414–2422. [CrossRef] [PubMed]
3. Delbecq, F.; Khodadadi, M.R.; Rodriguez Padron, D.; Varma, R.; Len, C. Isosorbide: Recent advances in catalytic production. *Mol. Catal.* **2020**, *482*, 110648. [CrossRef]
4. Tundo, P.; Aricò, F.; Gauthier, G.; Rossi, L.; Rosamilia, A.E.; Bevinakatti, H.S.; Sievert, R.L.; Newman, C.P. Green Synthesis of Dimethyl Isosorbide. *ChemSusChem* **2010**, *3*, 566–570. [CrossRef] [PubMed]
5. Rose, M.; Palkovits, R. Isosorbide as a Renewable Platform chemical for Versatile Applications—Quo Vadis? *ChemSusChem* **2012**, *5*, 167–176. [CrossRef]
6. Feng, X.; East, A.J.; Hammond, W.B.; Zhang, Y.; Jaffe, M. Overview of advances in sugar-based polymers. *Polym. Adv. Technol.* **2011**, *22*, 139–150. [CrossRef]
7. Fenouillot, F.; Rousseau, A.; Colomines, G.; Saint-Loup, R.; Pascault, J.P. Polymers from renewable 1,4:3,6-dianhydrohexitols (isosorbide, isomannide and isoidide): A review. *Prog. Polym. Sci.* **2010**, *35*, 578–622. [CrossRef]
8. Saxon, D.J.; Luke, A.M.; Sajjad, H.; Tolman, W.B.; Reineke, T.M. Next-generation polymers: Isosorbide as a renewable alternative. *Prog. Polym. Sci.* **2020**, *101*, 101196. [CrossRef]
9. Kricheldorf, H.R. "Sugar Diols" as Building Blocks of Polycondensates. *J. Macromol. Sci.* **1997**, *37*, 599–631. [CrossRef]
10. Cognet-Georjon, E.; Mechin, F.; Pascault, J.P. New polyurethanes based on 4,4′-diphenylmethane diisocyanate and 1,4:3,6 dianhydrosorbitol, 2. Synthesis and properties of segmented polyurethane elastomers. *Macromol. Chem. Phys.* **1996**, *197*, 3593–3612. [CrossRef]
11. Majdoub, M.; Loupy, A.; Fleche, G. Nouveaux polyéthers et polyesters à base d'isosorbide: Synthèse et caractérisation. *Eur. Polym. J.* **1994**, *30*, 1431–1437. [CrossRef]
12. Noordover, B.A.J.; van Staalduinen, V.G.; Duchateau, R.; Koning, C.E.; van, B.; Mak, M.; Heise, A.; Frissen, A.E.; van Haveren, J. Co- and Terpolyesters Based on Isosorbide and Succinic Acid for Coating Applications: Synthesis and Characterization. *Biomacromolecules* **2006**, *7*, 3406–3416. [CrossRef] [PubMed]
13. Thiyagarajan, S.; Gootjes, L.; Vogelzang, W.; van Haveren, J.; Lutz, M.; van Es, D.S. Renewable Rigid Diamines: Efficient, Stereospecific Synthesis of High Purity Isohexide Diamines. *ChemSusChem* **2011**, *4*, 1823–1829. [CrossRef]
14. Thiyagarajan, S.; Gootjes, L.; Vogelzang, W.; Wu, J.; van Haveren, J.; van Es, D.S. Chiral building blocks from biomass: 2,5-diamino-2,5-dideoxy-1,4-3,6-dianhydroiditol. *Tetrahedron* **2011**, *67*, 383–389. [CrossRef]
15. Thiyagarajan, S.; Wu, J.; Knoop, R.; Haveren, J.; Lutz, M.; Van Es, D. Isohexide hydroxy esters: Synthesis and application of a new class of biobased AB-type building blocks. *Rsc Adv.* **2014**, *4*, 7937–47950. [CrossRef]
16. Stensrud, K. Monoallyl, Monoglycidyl Ethers and Bisglycidyl Ethers of Isohexides. WO/2013/188253, 19 December 2013.
17. Torres, G.M.; Frauenlob, R.; Franke, R.; Börner, A. Production of alcohols via hydroformylation. *Catal. Sci. Technol.* **2015**, *5*, 34–54. [CrossRef]
18. Diebolt, O.; Müller, C.; Vogt, D. "On-water" rhodium-catalysed hydroformylation for the production of linear alcohols. *Catal. Sci. Technol.* **2012**, *2*, 773–777. [CrossRef]
19. Rodrigues, F.M.S.; Kucmierczyk, P.K.; Pineiro, M.; Jackstell, R.; Franke, R.; Pereira, M.M.; Beller, M. Dual Rh−Ru Catalysts for Reductive Hydroformylation of Olefins to Alcohols. *ChemSusChem* **2018**, *11*, 2310–2314. [CrossRef]
20. Furst, M.R.L.; Korkmaz, V.; Gaide, T.; Seidensticker, T.; Behr, A.; Vorholt, A.J. Tandem Reductive Hydroformylation of Castor Oil Derived Substrates and Catalyst Recycling by Selective Product Crystallization. *ChemCatChem* **2017**, *9*, 4319–4323. [CrossRef]
21. Takahashi, K.; Yamashita, M.; Nozaki, K. Tandem Hydroformylation/Hydrogenation of Alkenes to Normal Alcohols Using Rh/Ru Dual Catalyst or Ru Single Component Catalyst. *J. Am. Chem. Soc.* **2012**, *134*, 18746–18757. [CrossRef]
22. Fell, B.; Shanshool, J.; Asinger, F. Einstufige oxoalkoholsynthese mit einem Co2(CO)8/Fe(CO)5/N-methylpyrrolidin-katalysatorsystem. *J. Organomet. Chem.* **1971**, *33*, 69–72. [CrossRef]
23. Fell, B.; Geurts, A. Hydroformylierung mit Rhodiumcarbonyl-tert.-Amin-Komplexkatalysatoren. *Chem. Ing. Tech.* **1972**, *44*, 708–712. [CrossRef]
24. Cheung, L.L.W.; Vasapollo, G.; Alper, H. Synthesis of Alcohols via a Rhodium-Catalyzed Hydroformylation–Reduction Sequence using Tertiary Bidentate Amine Ligands. *Adv. Synth. Catal.* **2012**, *354*, 2019–2022. [CrossRef]

25. Mizoroki, T.; Kioka, M.; Suzuki, M.; Sakatani, S.; Okumura, A.; Maruya, K.-i. Behavior of Amine in Rhodium Complex–Tertiary Amine Catalyst System Active for Hydrogenation of Aldehyde under Oxo Reaction Conditions. *Bull. Chem. Soc. Jpn.* **1984**, *57*, 577–578. [CrossRef]
26. Hunter, D.L.; Moore, S.E.; Garrou, P.E.; Dubois, R.A. Dicyclopentadiene hydroformylation catalyzed by RhxCo4-x(CO)12 (x = 4, 2-0)/tertiary amine catalysts. *Appl. Catal.* **1985**, *19*, 259–273. [CrossRef]
27. Hunter, D.L.; Moore, S.E.; Dubois, R.A.; Garrou, P.E. Deactivation of rhodium hydroformylation catalysts on amine functionalized organic supports. *Appl. Catal.* **1985**, *19*, 275–285. [CrossRef]
28. Alvila, L.; Pakkanen, T.A.; Pakkanen, T.T.; Krause, O. Hydroformylation of olefins catalysed by rhodium and cobalt clusters supported on organic (Dowex) resins. *J. Mol. Catal.* **1992**, *71*, 281–290. [CrossRef]
29. Jurewicz, A.T.; Rollmann, L.D.; Whitehurst, D.D. Hydroformylation with Rhodium-Amine Complexes. Homogeneous Catalysis—II. *Adv. Chem.* **1974**, *132*, 240–251. [CrossRef]
30. Fuchs, S.; Lichte, D.; Dittmar, M.; Meier, G.; Strutz, H.; Behr, A.; Vorholt, A.J. Tertiary Amines as Ligands in a Four-Step Tandem Reaction of Hydroformylation and Hydrogenation: An Alternative Route to Industrial Diol Monomers. *ChemCatChem* **2017**, *9*, 1436–1441. [CrossRef]
31. Vanbésien, T.; Monflier, E.; Hapiot, F. Rhodium-catalyzed one pot synthesis of hydroxymethylated triglycerides. *Green Chem.* **2016**, *18*, 6687–6694. [CrossRef]
32. Rösler, T.; Ehmann, K.R.; Köhnke, K.; Leutzsch, M.; Wessel, N.; Vorholt, A.J.; Leitner, W. Reductive hydroformylation with a selective and highly active rhodium amine system. *J. Catal.* **2021**, *400*, 234–243. [CrossRef]
33. Gorbunov, D.; Nenasheva, M.; Naranov, E.; Maximov, A.; Rosenberg, E.; Karakhanov, E. Tandem hydroformylation/hydrogenation over novel immobilized Rh-containing catalysts based on tertiary amine-functionalized hybrid inorganic-organic materials. *Appl. Catal. A Gen.* **2021**, *623*, 118266. [CrossRef]
34. Hong, J.; Radojčić, D.; Ionescu, M.; Petrović, Z.S.; Eastwood, R. Advanced materials from corn: Isosorbide-based epoxy resins. *Polym. Chem.* **2014**, *5*, 5360. [CrossRef]
35. ÇAkmakÇI, E.; ŞEn, F.; Kahraman, M.V. Isosorbide Diallyl Based Antibacterial Thiol–Ene Photocured Coatings Containing Polymerizable Fluorous Quaternary Phosphonium Salt. *ACS Sustain. Chem. Eng.* **2019**, *7*, 10605–10615. [CrossRef]
36. Montassier, C.; Menezo, J.C.; Moukolo, J.; Naja, J.; Barbier, J.; Boitiaux, J.P. Furanic Derivatives Synthesis from Polyols by Heterogeneous Catalysis Over Metals. In *Studies in Surface Science and Catalysis*; Guisnet, M., Barrault, J., Bouchoule, C., Duprez, D., Pérot, G., Maurel, R., Montassier, C., Eds.; Elsevier: Amsterdam, The Netherlands, 1991; Volume 59, pp. 223–230.
37. Jiao, Y.; Torne, M.S.; Gracia, J.; Niemantsverdriet, J.W.; van Leeuwen, P.W.N.M. Ligand effects in rhodium-catalyzed hydroformylation with bisphosphines: Steric or electronic? *Catal. Sci. Technol.* **2017**, *7*, 1404–1414. [CrossRef]

Article

Vanadium(IV) Complexes with Methyl-Substituted 8-Hydroxyquinolines: Catalytic Potential in the Oxidation of Hydrocarbons and Alcohols with Peroxides and Biological Activity

Joanna Palion-Gazda [1], André Luz [2,3], Luis R. Raposo [2,3], Katarzyna Choroba [1], Jacek E. Nycz [1], Alina Bieńko [4], Agnieszka Lewińska [4], Karol Erfurt [5], Pedro V. Baptista [2,3], Barbara Machura [1,*], Alexandra R. Fernandes [2,3,*], Lidia S. Shul'pina [6], Nikolay S. Ikonnikov [6] and Georgiy B. Shul'pin [7,8,*]

Citation: Palion-Gazda, J.; Luz, A.; Raposo, L.R.; Choroba, K.; Nycz, J.E.; Bieńko, A.; Lewińska, A.; Erfurt, K.; V. Baptista, P.; Machura, B.; et al. Vanadium(IV) Complexes with Methyl-Substituted 8-Hydroxyquinolines: Catalytic Potential in the Oxidation of Hydrocarbons and Alcohols with Peroxides and Biological Activity. *Molecules* **2021**, *26*, 6364. https://doi.org/10.3390/molecules26216364

Academic Editor: Bartolo Gabriele

Received: 24 September 2021
Accepted: 18 October 2021
Published: 21 October 2021

Publisher's Note: MDPI stays neutral with regard to jurisdictional claims in published maps and institutional affiliations.

Copyright: © 2021 by the authors. Licensee MDPI, Basel, Switzerland. This article is an open access article distributed under the terms and conditions of the Creative Commons Attribution (CC BY) license (https://creativecommons.org/licenses/by/4.0/).

[1] Institute of Chemistry, University of Silesia, Szkolna 9, 40-006 Katowice, Poland; joanna.palion-gazda@us.edu.pl (J.P.-G.); katarzyna.choroba@us.edu.pl (K.C.); jacek.nycz@us.edu.pl (J.E.N.)
[2] Associate Laboratory i4HB-Institute for Health and Bioeconomy, NOVA School of Science and Technology, NOVA University Lisbon, 2819-516 Caparica, Portugal; af.luz@campus.fct.unl.pt (A.L.); l.raposo@campus.fct.unl.pt (L.R.R.); pmvb@fct.unl.pt (P.V.B.)
[3] UCIBIO—Applied Molecular Biosciences Unit, Department of Life Sciences, NOVA School of Science and Technology, NOVA University Lisbon, 2819-516 Caparica, Portugal
[4] Faculty of Chemistry, University of Wroclaw, F. Joliot-Curie 14, 50-383 Wroclaw, Poland; alina.bienko@chem.uni.wroc.pl (A.B.); agnieszka.lewinska@chem.uni.wroc.pl (A.L.)
[5] Department of Chemical Organic Technology and Petrochemistry, Silesian University of Technology, Krzywoustego 4, 44-100 Gliwice, Poland; karol.erfurt@polsl.pl
[6] A.N. Nesmeyanov Institute of Organoelement Compounds, Russian Academy of Sciences, Ulitsa Vavilova 28, 119991 Moscow, Russia; shulpina@ineos.ac.ru (L.S.S.); ikonns@ineos.ac.ru (N.S.I.)
[7] N.N Semenov Federal Research Center for Chemical Physics, Russian Academy of Sciences, Ulitsa Kosygina 4, 119991 Moscow, Russia
[8] Chair of Chemistry and Physics, Plekhanov Russian University of Economics, Stremyannyi Pereulok 36, 117997 Moscow, Russia
* Correspondence: barbara.machura@us.edu.pl (B.M.); ma.fernandes@fct.unl.pt (A.R.F.); gbsh@mail.ru (G.B.S)

Abstract: Methyl-substituted 8-hydroxyquinolines (Hquin) were successfully used to synthetize five-coordinated oxovanadium(IV) complexes: [VO(2,6-(Me)$_2$-quin)$_2$] (**1**), [VO(2,5-(Me)$_2$-quin)$_2$] (**2**) and [VO(2-Me-quin)$_2$] (**3**). Complexes **1**–**3** demonstrated high catalytic activity in the oxidation of hydrocarbons with H$_2$O$_2$ in acetonitrile at 50 °C, in the presence of 2-pyrazinecarboxylic acid (PCA) as a cocatalyst. The maximum yield of cyclohexane oxidation products attained was 48%, which is high in the case of the oxidation of saturated hydrocarbons. The reaction leads to the formation of a mixture of cyclohexyl hydroperoxide, cyclohexanol and cyclohexanone. When triphenylphosphine is added, cyclohexyl hydroperoxide is completely converted to cyclohexanol. Consideration of the regio- and bond-selectivity in the oxidation of n-heptane and methylcyclohexane, respectively, indicates that the oxidation proceeds with the participation of free hydroxyl radicals. The complexes show moderate activity in the oxidation of alcohols. Complexes **1** and **2** reduce the viability of colorectal (HCT116) and ovarian (A2780) carcinoma cell lines and of normal dermal fibroblasts without showing a specific selectivity for cancer cell lines. Complex **3** on the other hand, shows a higher cytotoxicity in a colorectal carcinoma cell line (HCT116), a lower cytotoxicity towards normal dermal fibroblasts and no effect in an ovarian carcinoma cell line (order of magnitude HCT116 > fibroblasts > A2780).

Keywords: vanadium(IV) complexes; biological activity; catalytic properties; 8-hydroxyquinoline; cytotoxicity studies

1. Introduction

In the last three decades, vanadium coordination compounds have received increasing interest due to their structural features [1–44], catalytic applications [21–33,45,46] and medicinal importance [33–44,47–63]. Particular attention has been paid to bis-chelated $V^{IV}O$ and V^VO complexes of the general formula $[V^{IV}O(N\cap O)_2]$, $[V^{IV}O(N\cap O)_2(solvent)]$ and $[V^VO(OR)(N\cap O)_2]$. Among them there is the vanadium maltolate complex [VO(ethylm altolate)2] which entered phase IIa clinical trials as an antidiabetic agent [56]. For $V^{IV}O$ complexes bearing bidentate picolinate ligands, it was found that their pharmacological potential is strongly dependent on the structural modification of organic ligands and ligand arrangement around the metal center [54,57–61]. Exemplarily, the introduction of an electron-withdrawing halogen atom or electron-donating alkyl group at the fifth or third position of the pyridine ring leads to stronger insulin-enhancing activity of $[VO(picolinate)_2]$ and $[VO(picolinate)_2(solvent)]$ in comparison with $[VO(picolinate)_2(H_2O)]$. On the contrary, investigations of the substituent effect on the antiproliferative potential of vanadium complexes $[V^VO(OMe)(N\cap O)_2]$ bearing 8-hydroxyquinoline ligands on HCT116 and A2780 cancer cell lines, showed that the introduction of substituents into the 8-hydroxyquinoline backbone at 5- and 5,7-positions induces a reduction of the antiproliferative effect in relation to $[VO(OMe)(quin)_2]$, and the complexes $[V^VO(OMe)(N\cap O)_2]$ with hydroxyquinoline (quin) substituted only in the 5-position were more cytotoxic than those with substituents in the 5,7-positions of the quin backbone. Nevertheless, as all the vanadium(V) complexes $[V^VO(OR)(N\cap O)_2]$ with 8-hydroxyquinoline derivatives showed a significantly lower IC_{50} towards the A2780 cell line, other than oxovanadium and dioxovanadium complexes previously reported [62], these systems deserve further intensive studies. What is also important, is that monomeric oxidovanadium(V) complexes $[VO(OMe)(N\cap O)_2]$ with nitro- or halogen-substituted quinolin-8-olate ligands were found to be very promising in view of their catalytic properties. These complexes exhibit high catalytic activity toward the oxidation of inert alkanes to alkyl hydroperoxides by H_2O_2 in aqueous acetonitrile, with the yield of oxygenate products up to 39% and a TON of 1730 for 1 h [63].

In continuation of our studies on monomeric oxovanadium complexes with bidentate monoanionic ligands [62,63], we present herein the biological and catalytic potential of three oxovanadium(IV) complexes with methyl-substituted 8-hydroxyquinolines, they are $[VO(2,6-(Me)_2-quin)_2]$ (**1**), $[VO(2,5-(Me)_2-quin)_2]$ (**2**) and $[VO(2-Me-quin)_2]$ (**3**). Two of them (**1** and **2**) have been obtained for the first time, while compound **3**, the X-ray structure of which has been presented previously [64], was included to obtain more reliable structure–activity relationships for these systems. In contrast to the previously reported $[V^VO(OMe)(N\cap O)_2]$ with 8-hydroxyquinoline derivatives bearing substituents in the 5- or 5,7-positions of the quin backbone, all the complexes here presented contain a vanadium(IV) ion and they are five-coordinated, which was evidenced by the X-ray diffraction analysis, EPR and UV–Vis spectroscopy. The catalytic potential of complexes **1–3** was examined for the oxidation of alkanes with H_2O_2 and compared with the activity of the classic system vanadate ion plus pyrazine-carboxylic acid (PCA). To evaluate the antiproliferative effect of complexes **1–3**, HCT116 and A2780 cancer cell lines, and normal dermal fibroblasts were used.

2. Results and Discussion

2.1. Synthesis

To synthetize the oxidovanadium complexes with methyl-substituted 8-hydroxyquinolines, $[VO(2,6-(Me)_2-quin)_2]$ (**1**), $[VO(2,5-(Me)_2-quin)_2]$ (**2**), the previously reported procedure based on the reaction of bis(acetylacetonato)oxidovanadium(IV) with the corresponding 8-hydroxyquinoline derivative in open air was employed [62,63]. Remarkably, in the reaction with the use of quinH, 5-Cl-quinH, 5-NO_2-quinH, 5,7-Cl_2-quinH, 5,7-$(Me)_2$-quinH, 5,7-Cl,I-quinH and 5,7-I_2-quinH, 5,7-$(Me)_2$-quinH, the vanadium(IV) of the starting $[VO(acac)_2]$ undergoes oxidation by molecular oxygen and the acetylacetonato ligands are exchanged by the corresponding quinolin-8-olate ions to give six-coordinated $[V^VO(OMe)(N\cap O)_2]$. In contrast, the reactions of $[VO(acac)_2]$ with 2-Me-quinH, 2,6-$(Me)_2$-quinH and 2,5-$(Me)_2$-quinH resulted in the formation of five-coordinated $[V^{IV}O(N\cap O)_2]$. It indicates that the methyl-functionalization of quinH strengthen its coordination capacity to the V(IV) ion, and steric hindrance induced by the methyl group at the 2-position in the pyridine ring facilitates the formation of five-coordinated complexes.

2.2. Molecular Structure

Perspective views of the molecular structures of **1** and **2** together with the atom numbering are depicted in Figure 1. The atoms V(1) and O(2) in the structure **1** are located on a twofold crystallographic axis, thus the molecule [VO(5,6-(Me)$_2$-quin)$_2$] has crystallographically-imposed twofold symmetry.

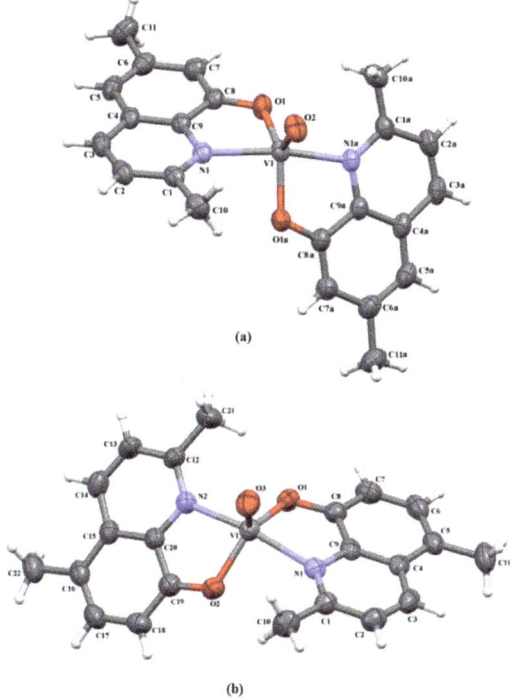

Figure 1. The molecular structures of **1** [symmetry code: (a) = −x, y, 1/2 − z] (a) and **2** (b). Displacement ellipsoids are drawn at the 50% probability level.

The five-coordinated vanadium(IV) ion adopts a coordination geometry lying between square pyramidal and trigonal bipyramidal. The angular structural index parameter τ [65], expressed as the difference between the two largest angles divided by 60, has a value of 0.57 for **1** and 0.56 for **2**. The atom V(1) is shifted ~0.6 Å towards the oxo ion from the least-squares plane defined by the nitrogen and oxygen atoms of two quinolin-8-olate bidentate ligands. The bite angle of the chelating ligand is ~80° (Tables S1–S3, Supplementary Materials), and the dihedral angle between the least-squares planes formed by the organic ligands is equal to 60.38(4)° for **1** and 50.32(3)° for **2**. This kind of coordination geometry seems to be typical for [VO(N∩O)$_2$] bearing two N,O-donor bidentate ligands, contrary to [VO(O∩N∩N∩O)] of tetradentate Schiff base ligands in which five-coordinated vanadium(IV) generally displays a distorted square-pyramidal environment with a basal square plane constituted by the two iminic nitrogen and two phenoxide oxygen atoms [4,6,64,66–82] (Table S7, Supplementary Materials). The V = O [1.590(2) and 1.595(3) Å], V–O [1.921(1) and 1.924(2) Å] and V–N [2.122(2) and 2.125(2) Å] bond lengths in **1** and **2** (Tables S1 and S2) are in good agreement with those reported for five-coordinated vanadyl compounds (Tables S3 and S7, Supplementary Materials). More detailed structural parameters of the designed complexes are included in Tables S4–S6 and Figures S3–S8 in Supplementary Materials.

The phase purity of **1** and **2** was evidenced by comparing the powder X-ray diffraction (PXRD) patterns of the powdery sample with those generated by simulation based on single-crystal structures. As shown in Figure S9, the bulk powder samples give patterns consistent with those obtained theoretically from the single-crystal structure.

2.3. EPR Spectroscopy

The oxidation state of vanadium in complexes **1–3** was confirmed by the EPR spectra (Figure 2). The X- and Q-band EPR powder spectra of all complexes were recorded at room temperature. The X-band spectrum shows intense central field broad structured bands with no detectable hyperfine structure at g = 1.982 for **1**, 1.986 for **2** and 1.996 for **3**. The Q-band spectrum presents slightly resolved peaks, and the nature of the peaks is the same as in the X-band. The 77 K powder EPR spectra recorded at the X-band show that there is a slight increase in the signal intensity.

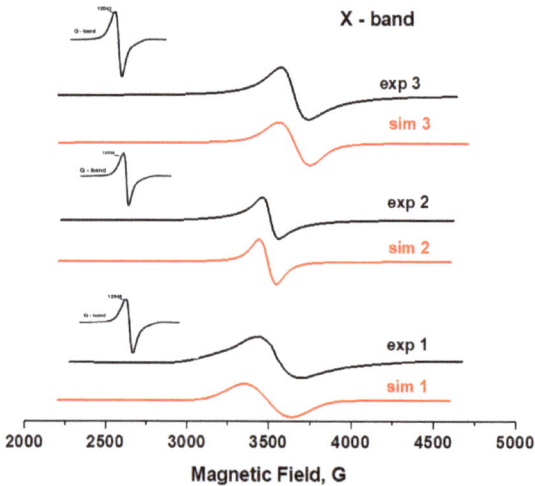

Figure 2. The X-band EPR spectra of **1–3** at 77 K together with the spectrum calculated by computer simulation of the experimental spectra with spin Hamiltonian parameters given in the text.

EPR frozen solution spectra of compounds **1–3** in DMSO (Figure 3) and CH$_3$CN (Figure S10) show eight lines of hyperfine splitting of a parallel and perpendicular orientation, proving the interaction of S = 1/2 with the nucleus spin of one vanadium and hence the formation of mononuclear compounds. The spectra of these mononuclear compounds may be simulated using the spin Hamiltonian parameters g_x = g_y = 1.979, g_z = 1.942, A_x = A_y = 65 G, A_z = 180 G for **1**, g_x = g_y = 1.977, g_z = 1.958, A_x = A_y = 52 G, A_z = 169 G for **2**, and g_x = g_y = 1.994, g_z = 1.946, A_x = A_y = 50 G, A_z = 169G for **3**, respectively, which are typical for oxidovanadium(IV) compounds with an analogous N$_2$O$_2$ donor set of the ligands in the vanadium xy plane [38,83,84], and in agreement with the molecular structure determined by X-ray crystal structure studies of **1–3**.

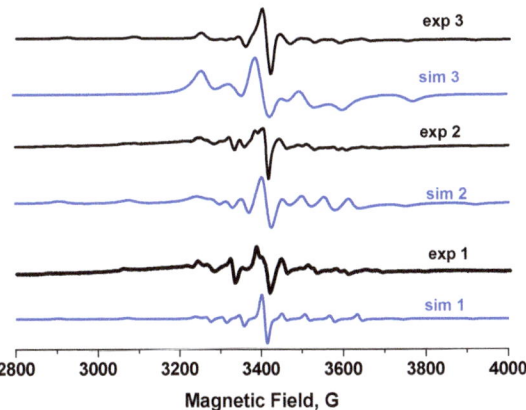

Figure 3. EPR frozen solution spectra (at 77 K) of compounds **1–3**; in aqueous 2% DMSO.

In order to investigate the stability of the studied complexes, EPR spectra were recorded depending on the time (Figure 4). The constant position of bands confirms the stability of the V(IV) complexes in the solution.

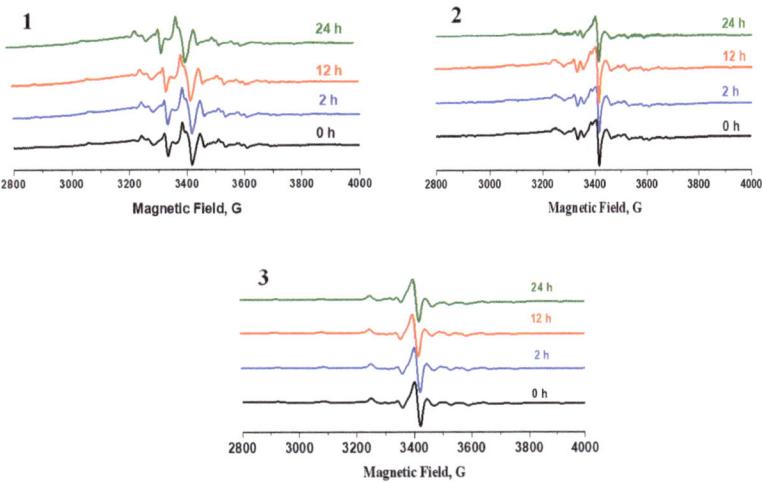

Figure 4. EPR stability spectra (frozen solution at 77 K in aqueous 2% DMSO) of compounds **1–3**. Spectra were recorded every for 24 h.

2.4. Absorption Spectroscopy

The yellowish-brown tetravalent d^1 complexes **1** and **3** display three ligand-field transitions $d_{xy} \to d_{xz}, d_{yz}$ and $d_{xy} \to d_{x^2-y^2}$ and $d_{xy} \to d_{z^2}$ at 701, 585 and 523 nm for **1** and 744, 572 and 501 nm for **3**. In the case of **2**, only two bands of low intensity in the range 500–850 nm, are observed. An expected a third transition attributed to $d_{xy} \to d_{z^2}$ is most likely masked by the intense ligand-metal charge-transfer (LMCT) transition with a maximum at 398 nm. In the spectra of **1** and **3**, the absorptions assigned transitions from the p_π orbital of the phenolate oxygen to d_π orbitals of the vanadium center (LMCT) occur at 380 and 383 nm, respectively. The higher energy bands of **1–3**, are attributable to the spin-allowed ligand centered transitions (IL) $\pi \to \pi^*$ of quinolin-8-olate (Figure S11, Supplementary Materials).

UV–Vis spectroscopy was also used to study the stability of the five-coordinated oxidovanadium(IV) complexes in solution, and UV–Vis spectra of **1–3** in DMSO and CH$_3$CN (10^{-4} M) were collected once every two hours over 24 h at room temperature. As observed, the main bands remained constant in the electronic spectra, indicating stability of the V(IV) complexes in solution (Figures 5 and S12, Supplementary Materials).

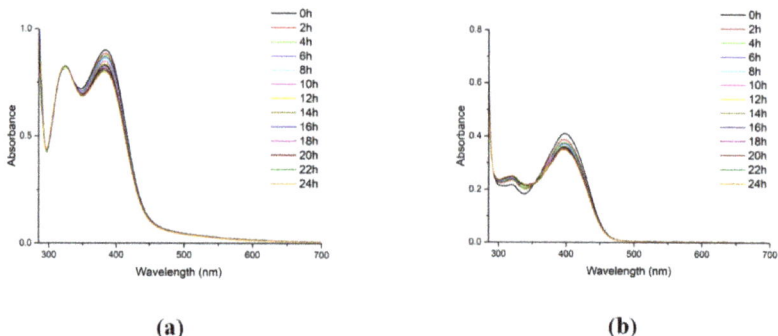

Figure 5. UV–Vis stability spectra in DMSO (10^{-4} M) for compounds **1** (a) and **2** (b). Spectra were recorded every 2 h for 24 h.

2.5. Catalytic Oxidations with Hydrogen Peroxide

One of us discovered in 1993, a system for the efficient oxidation of various organic compounds with hydrogen peroxide, based on a simple vanadate ion compound. The obligatory component of this system was PCA (pyrazinecarboxylic acid). This reaction takes place at low temperatures in a solution of acetonitrile [85]. Further, this system was studied in detail, including the oxidation of alkanes, olefins, arenes, and alcohols with hydrogen peroxide and other oxidizing agents [86–88]. In the absence of a catalyst, the reaction proceeds extremely slowly, in five hours the yield of products is no more than 3–5% [89–92].

In this work, we present a study of the catalytic activity of **1–3** and the effect of 2-pyrazinecarboxylic acid on the activities of these complexes. It turned out that all complexes exhibit high catalytic activity in the oxidation of alkanes, but only in the presence of PCA. The curves for the accumulation of oxidation products are shown in Figures 6–9.

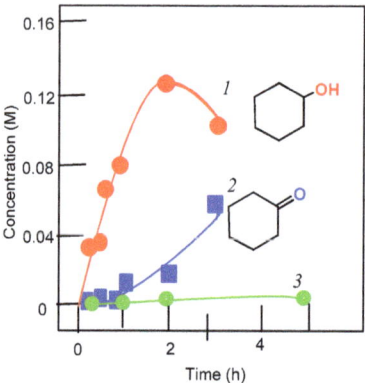

Figure 6. Oxidation of cyclohexane to cyclohexanol (curves 1 and 3) and cyclohexanone (curve 2) with hydrogen peroxide catalyzed by compound **1** in the presence of PCA (curves 1 and 2) and in the absence of PCA (curve 3). Conditions: cyclohexane (0.46 M); H$_2$O$_2$ (2.0 M, 50% aqueous); complex **1** (5×10^{-4} M); PCA (2×10^{-3} M) in MeCN at 50 °C. Concentrations of cyclohexanone and cyclohexanol were measured after reduction of the aliquots with solid PPh$_3$.

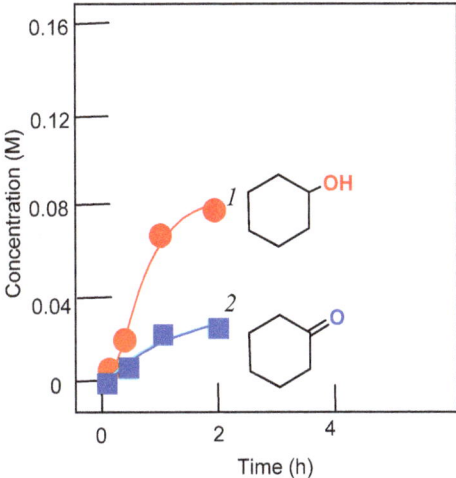

Figure 7. Oxidation of cyclohexane to cyclohexanol (curves 1) and cyclohexanone (curve 2) with hydrogen peroxide catalyzed by compound **1** in the presence of HNO$_3$. Conditions: cyclohexane (0.46 M); H$_2$O$_2$ (2.0 M, 50% aqueous); complex **1** (5 × 10^{-4} M); HNO$_3$ (0.05 M) in MeCN at 50 °C. Concentrations of cyclohexanone and cyclohexanol were measured after reduction of the aliquots with solid PPh$_3$.

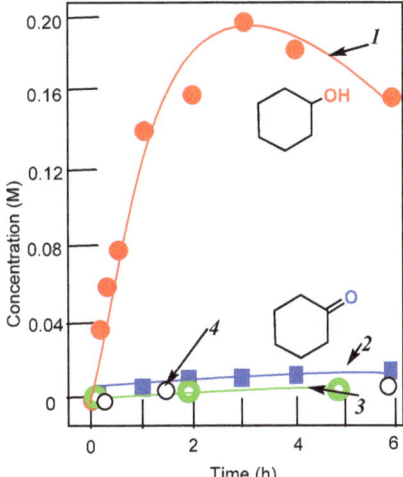

Figure 8. Oxidation of cyclohexane to cyclohexanol (curves 1 and 3) and cyclohexanone (curve 2) with hydrogen peroxide catalyzed by compound **2** in the presence of PCA (curves 1 and 2) and in the absence of PCA (curve 3); symbols marked by the number 4 show the reaction carried out in the absence of PCA and HNO$_3$. Conditions: cyclohexane (0.46 M); H$_2$O$_2$ (2.0 M, 50% aqueous); complex **2** (5 × 10^{-4} M); PCA (2 × 10^{-3} M), HNO$_3$ (0.05 M) in MeCN at 50 °C. Concentrations of cyclohexanone and cyclohexanol were measured after reduction of the aliquots with solid PPh$_3$.

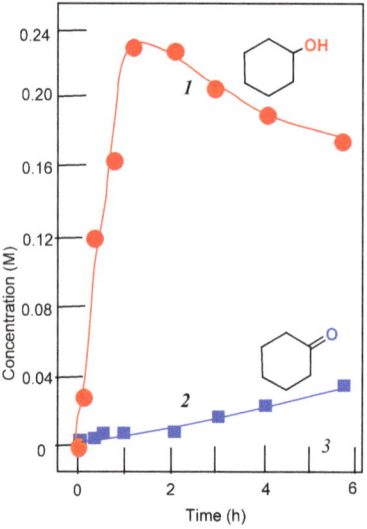

Figure 9. Accumulation of cyclohexanol (curve 1) and cyclohexanone (curve 2) in the oxidation of cyclohexane (0.46 M) with hydrogen peroxide (2.0 M, 50% aqueous) catalyzed by compound **3** (5×10^{-4} M); PCA (2×10^{-3} M) in MeCN at 50 °C. Concentrations of cyclohexanone and cyclohexanol were measured after reduction of the aliquots with solid PPh$_3$. The yield of cyclohexane oxidation products was 48%.

In the case of using complexes **1–3**, the yields of alkane oxidation products are noticeably higher than in reactions catalyzed by the vanadate anion-PCA system (Table S8). Compared to other vanadium complexes, that we and other researchers have used in the oxidative catalysis of alkanes and alcohols, we can say that the complexes described in this article exhibit a high activity. We have also studied the oxidation of cyclohexane in the presence of nitric acid. In this case, for complex **1**, the yield of oxidation products is much lower than in the reaction using PCA (see Figure 7). And in the case of complexes **2** and **3**, the oxidation reactions of cyclohexane do not lead to any noticeable yields of oxidation products.

Consideration of the regio- and bond-selectivity in the oxidation of n-heptane and methylcyclohexane indicates that the oxidation proceeds with the participation of free hydroxyl radicals. The regio-selectivity parameters for the oxidation of n-heptane were obtained for complex **2**: C (1):C (2):C (3):C (4) = 1.0:5.0:5.0:4.4. The bond-selectivity parameters for the oxidation of methylcyclohexane were also obtained: for complex **1**: 1°:2°:3° = 1.0:6.3:16.7; for complex **2**: 1°:2°:3° = 1.0:5.5:15.0; for complex **3**: 1°:2°:3° = 1.0:5.7:16.0, respectively. These values are close to the parameters obtained for oxidation reactions with the participation of hydroxyl radicals, although they are somewhat higher, apparently due to steric hindrances created by chelating ligands around the catalytic center in the molecules of our complexes [93].

The complexes show moderate activity in the oxidation of alcohols. The yields for the oxidation of phenylethanol to acetophenone with tert-butyl hydroperoxide under catalysis with complexes **1–3**, were 46%, 23% and 32%, respectively, at a temperature of 50 °C, in acetonitrile for four hours. Hydrogen peroxide is much less productive in these reactions. In analogous oxidation reactions of cyclohexanol to cyclohexanone, corresponding yields were 10%, 5.5% and 5.5% after 5 h. In the oxidation of cyclohexene, yields of cyclohexene-1-ol were 10%, 6.5% and 5.5% correspondingly for complexes **1–3**. Epoxide and other products of oxidation were formed in insignificant quantities.

2.6. Viability Studies

HCT116 and A2780 cancer cell lines and normal dermal fibroblasts were used to evaluate the antiproliferative effect of complexes 1–3 (Figures 10 and S13 and Table 1).

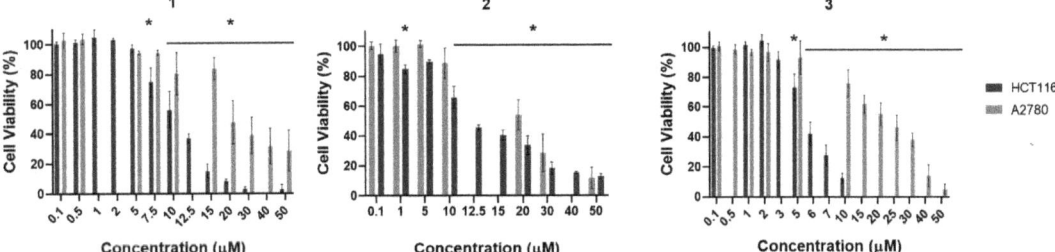

Figure 10. Antiproliferative effect of complexes 1–3 in the HCT116 and A2780 cancer cell lines, after 48 h, evaluated by the MTS method. Cell viabilities were normalized to DMSO 0.1% (v/v) (vehicle control). The results presented are mean ± standard deviation of three independent assays. An asterisk indicates a p-value inferior to 0.05.

Table 1. Relative IC_{50} values of complexes 1–3 in HCT116 and A2780 cancer cell lines and in normal dermal fibroblasts.

Cell Lines	1 (μM)	2 (μM)	3 (μM)
HCT116	12.5 ± 0.63	9.6 ± 0.54	5.9 ± 0.66
A2780	17.5 ± 1.13	20.7 ± 3.11	50.9 ± 2.75
Fibroblasts	15.8 ± 1.49	8.1 ± 2.75	14.5 ± 2.33

Our results show that complex 1 displayed similar cytotoxicity in both cancer cell lines and in normal dermal fibroblasts (Figures 10 and S13 and Table 1). Complex 2 displayed a higher cytotoxicity in normal dermal fibroblasts compared to HCT116 and A2780 tumor cell lines (Figures 10 and S13 and Table 1). On the other hand, complex 3 showed a higher cytotoxicity in the HCT116 cell line (5.9 μM) compared to normal dermal fibroblasts and the A2780 tumor cell line (Figures 11 and S13 and Table 1). Moreover, HCT116 was the cancer cell model more sensitive to the complexes (Figures 10 and S13 and Table 1). Interestingly, all complexes show a higher cytotoxicity (lower IC_{50}) compared to cisplatin (IC_{50} of 15 μM; see Supplementary Figure S13).

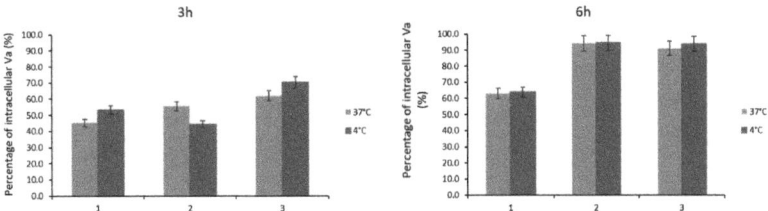

Figure 11. Internalization of complexes evaluated by the determination of the amount of vanadium (determined by ICP-AES) present in the cellular fraction of HCT116 after exposure of HCT116 cells to 20 × IC_{50} concentrations of complexes 1–3 for 3 and 6 h at 4 °C and 37 °C. The results presented are mean ± standard deviation of three independent assays.

The complexes herein described exhibit an IC_{50} in the range of 5.9 to 50.9 μM in HCT116 and A2780 cancer cell lines (Figures 11 and S13 and Table 1). Previously published complexes bearing [VO(OMe)(quin)$_2$] backbones, possess an IC_{50} between 0.96 and 10.09 μM in HCT116 and A2780 cancer cell lines and from 11.24 to over 100 μM in normal dermal fibroblasts [62]. This may be an indication that complexes bearing VO(2-Me-Quin) backbones are not as promising for therapeutic use as previously described complexes bearing [VO(OMe)(quin)$_2$] backbones [62]. Published results of the IC$_{50}$ values of complexes bearing hydroxyquinoline derived ligands in cancer cell lines other than HCT116 and A2780 range from 0.9 to 219 µM, and oxovanadium and dioxidovanadium complexes present IC$_{50}$ values between 0.96 to 224.5 µM, which indicates that the complexes here described have an IC$_{50}$ in the lower end of the range (Table 1) [62,94,95].

To further understand the mechanisms of cytotoxicity associated with exposure to complexes **1–3**, additional biological studies in the HCT116 cell line were carried out.

2.7. Complex Internalization

Internalization of the complexes was evaluated by exposing HCT116 cancer cells to 20 × IC$_{50}$ concentrations of the complexes **1–3** for 3 and 6 h at 4 °C and 37 °C (Figure 11).

Our results show that temperature does induce statistically significant alterations in the amount of complex internalized, which suggests that there is no active transport of complexes into cells after 3 or 6 h (Figure 11), which has also been described in the literature with oxovanadium complexes [96]. After 6 h incubation, approximately 94% and 91% of complexes **2** and **3**, respectively, and 60% of complex **1** were found in the HCT116 cellular fraction (Figure 11). These results suggest that the cytotoxicity might be directly correlated with the amount of internalized complex over time (Figures 10 and 11) and complexes **2** and **3** appear to be retained within HCT116 cells (Figure 11).

2.8. Induction of Apoptosis in the HCT116 Cell Line Exposed to Complexes **1–3**

To understand if the loss of cellular viability induced by exposure to complexes **1–3** is associated with the triggering of programed cell death, HCT116 cells were exposed to the respective IC$_{50}$ concentrations for 48 h, and the Annexin V-Alexa fluor 488/PI dead cell apoptosis assay (ThermoFisher, Waltham, MA, USA) was used to determine the percentage of apoptotic cells (Figure 12).

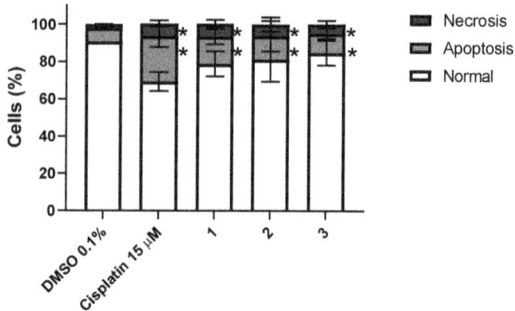

Figure 12. Apoptosis induction in the HCT116 cell line evaluated by flow cytometry after 48 h exposure to IC$_{50}$ concentrations of complexes **1–3**. DMSO 0.1% (v/v) was used as a negative control and Cisplatin 15 µM was used as a positive control. The results presented are mean ± standard deviation of three independent assays. An asterisk indicates a p-value inferior to 0.05.

Our results show a statistically significant increase in the percentage of apoptotic cells due to exposure to complexes **1–3** (Figure 12). HCT116 cells displayed 14.5%, 12.6% and 10.4% of apoptotic cells after 48 h exposure to complexes **1–3**, respectively, while the DMSO condition presented only 7.3% of apoptotic cells (Figure 12). There is also an increase in necrotic cells when compared with the DMSO control, which has 2% of necrotic cells, in cells exposed to complexes **1–3**, of 7.0%, 6.2% and 5.0%, respectively. All complexes induced lower apoptotic cell death when compared with cisplatin (Figure 12). The induction of apoptosis has been reported for other vanadium complexes bearing 8-hydroxyquinoline-based ligands and dioxidovanadium(V) complexes [62,95]. Based on the viability data (Figures 10 and S13 and Table 1) and internalization data (Figure 11) we were expecting a higher cell death for complex **2** and particularly for complex **3**. Since we did not observe this trend, additional cell death mechanisms might be involved in the loss of cell viability. To further explain this issue, we assessed autophagic HCT116 cell death after exposure to complexes **1–3**.

*2.9. Induction of Autophagy in the HCT116 Cell Line Exposed to Complexes **1–3***

Another mechanism of cell death triggered by exposure to metallic complexes is autophagy. The Autophagy Assay kit (Abcam) was used to determine the percentage of autophagic cells present after 48 h exposure to IC_{50} concentrations of complexes **1–3** (Figure 13).

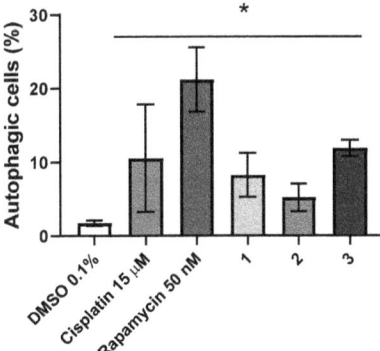

Figure 13. Autophagy induction in the HCT116 cell line evaluated by flow cytometry after 48 h exposure to IC_{50} concentrations of complexes **1–3**. DMSO 0.1% (v/v) was used as a negative control and Cisplatin 15 µM and rapamycin 50 nM were used as positive controls. The results presented are mean ± standard deviation of three independent assays. An asterisk indicates a *p*-value inferior to 0.05.

Exposure to complexes **1–3** led to a 7×, 4× and 10× increase of autophagic cells, respectively when compared to the DMSO control (Figure 13). Interestingly, complex **3** was able to induce more autophagic cell death compared to cisplatin (but to a lesser extent compared to rapamycin) (Figure 13). Therefore, induction of apoptosis and autophagy are responsible for the total loss of cell viability induced by the complexes in HCT116 cells (Figures 10, 11 and 13 and Table 1).

*2.10. Intracellular Reactive Oxygen Species (ROS) Production in the HCT116 Cell Line Exposed to Complexes **1–3***

ROS are known triggers of apoptosis and autophagy in cells exposed to metallic complexes. ROS production was evaluated in HCT116 cells exposed to IC_{50} concentrations of complexes **1–3** for 48 h (Figure 14).

Our results show a statistically significant increase of 3.3× and 2.3× of intracellular ROS in HCT116 cells exposed to complexes **2** and **3**, respectively when compared to the DMSO control, although lower than cisplatin (Figure 14). Although not statistically significant, complex **1** also induces a 1.15× increase in the amount of intracellular ROS in HCT116 cells (Figure 14).

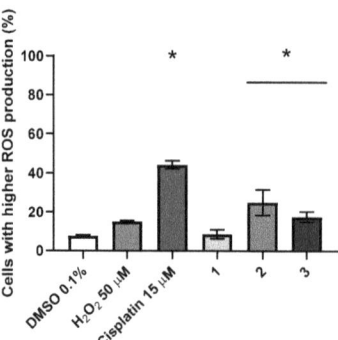

Figure 14. Intracellular ROS production in the HCT116 cell line after 48 h exposure to IC$_{50}$ concentrations of complexes **1–3**, evaluated by flow cytometry. DMSO 0.1% (v/v) was used as a negative control and Cisplatin 15 µM and H$_2$O$_2$ 50 µM were used as positive controls. The results presented are mean ± standard deviation of three independent assays. An asterisk indicates a p-value inferior to 0.05.

2.11. Evaluation of Alterations in the Mitochondrial Membrane Potential of HCT116 Cells Exposed to Complexes **1–3**

To evaluate if the observed apoptosis induction in the HCT116 cell line is triggered by the intrinsic pathway, due to destabilization of mitochondria, changes in the mitochondrial membrane potential were evaluated using JC-1 dye (Abnova). This is a green fluorescent dye as a monomer; however, it will aggregate in the presence of a normal mitochondrial potential, which will cause a red-shift in emission spectra [97]. Measuring the green and red fluorescence by flow cytometry it is possible to calculate normalized ratios that can show us if there is an increase or decrease in the mitochondrial membrane potential (Figure 15).

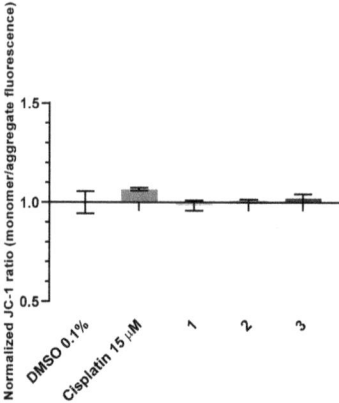

Figure 15. Alterations of the mitochondrial membrane potential in the HCT116 cell line after 48 h exposure to IC$_{50}$ concentrations of complexes **1–3**, evaluated by flow cytometry. DMSO 0.1% (v/v) was used as a negative control and Cisplatin 15 µM was used as a positive control. The results presented are mean ± standard deviation of three independent assays.

Our results shown that there is no significant change of the mitochondrial membrane potential in HCT116 cells exposed to complexes **1–3** when compared with the DMSO control (Figure 15). The cisplatin control displays a small increase in the JC-1 (monomer/aggregate) ratio, which is indicative of mitochondrial membrane destabilization, although not statistically significant (Figure 15).

3. Materials and Methods

3.1. Materials

2-methyl-8-hydroxyquinoline was commercially available (Sigma Aldrich, Darmstadt, Germany), while 2,5-dimethyl-8-hydroxyquinoline and 2,6-dimethyl-8-hydroxyquinoline were synthesized according to procedures described previously [98]. The analytical data (FT-IR and HR-MS spectra, X-ray analysis and PXRD spectrum) for [VO(2-Me-quin)$_2$] (3) provided a good agreement with those reported in reference [64].

3.2. Synthesis of [VO(2,6-(Me)$_2$-quin)$_2$] (1) and [VO(2,5-(Me)$_2$-quin)$_2$] (2)

A mixture of VO(acac)$_2$ (1 mmol, 0.26 g) and the appropriate 2,5-dimethyl-8-hydroxyquinoline or 2,6-dimethyl-8-hydroxyquinoline (2 mmol; 0.35 g) was suspended in toluene (30 mL), and the resulting solution was refluxed for 2 h in open air. After several days the crystalline solid of the vanadium(IV) complex was collected. The crystals suitable for X-ray analysis were obtained by recrystallization from an acetonitrile/chloroform mixture (1/1 v/v).

[VO(2,6-(Me)$_2$-quin)$_2$] (1): Yield 55%. HR-MS (ESI): calcd for C$_{22}$H$_{20}$N$_2$O$_3$NaV$^+$ [M + Na]$^+$ 434.0811, found 434.0818. Anal. Calc. for C$_{22}$H$_{20}$N$_2$O$_3$V (411.34 g/mol): C 64.25, H 4.90, N 6.81%. Found: C 64.55, H 4.73, N 6.78%. IR (KBr, cm^{-1}): 3436(s), 3028(w), 2920(m), 1613(w), 1575(s), 1496(s), 1469(m), 1439(w), 1401(s), 1377(w), 1346(m), 1266(s), 1208(w), 1041(s), 1132(s), 1033(w), 987(m), 976 (m), 946(s), 847(m), 789(w), 782(w), 725(m), 686(m), 653(m), 634(m), 618(w), 600(w), 533(m), 432(w). UV–Vis (DMSO, λ_{max}, nm (ε, dm^3·mol^{-1}·cm^{-1})): 268 (25170), 312 (1930), 380 (2310), 523 (530), 585 (280), 701 (110).

[VO(2,5-(Me)$_2$-quin)$_2$] (2): Yield 75%. HRMS (ESI): calcd for C$_{22}$H$_{20}$N$_2$O$_3$NaV$^+$ [M + Na]$^+$ 434.0811, found 434.0813. Anal. Calc. for C$_{22}$H$_{20}$N$_2$O$_3$V (411.34 g/mol): C 64.25, H 4.90, N 6.81%. Found: C 63.94, H 4.69, N 6.52%. IR (KBr, cm^{-1}): 3438(m), 3028(m), 2892(w), 2861(w), 1597(m), 1571(s), 1515(s), 1462(s), 1434(s), 1409(m), 1374(m), 1314(s), 1291(m), 1263(s), 1245(m), 1223(w), 1157(m), 1093(s), 983(s), 947(w), 836(s), 819(w), 787(m), 767(s), 747(m), 644(m), 632(m), 523(m), 501(w), 452(w). UV–Vis (DMSO, λ_{max}, nm (ε, dm^3·mol^{-1}·cm^{-1})): 320 (3960), 398 (6600), 582 (100), 753 (80).

3.3. X-ray Crystal Structure Determination

The X-ray diffraction data for complexes 1 and 2 were collected using an Oxford Diffraction four-circle diffractometer Gemini A Ultra with an Atlas CCD detector using graphite monochromated MoKα radiation (λ = 0.71073 Å) at room temperature. Diffraction data collection, cell refinement and data reduction were performed using the CrysAlis^{Pro} software [99]. The structures were solved by direct methods using SHELXS and refined by full-matrix least-squares on F^2 using SHELXL-2014 [100]. All the non-hydrogen atoms were refined anisotropically, and hydrogen atoms were placed in calculated positions and refined with riding constraints: d(C–H) = 0.93 Å, U_{iso}(H) = 1.2 U_{eq}(C) (for aromatic) and d(C–H) = 0.96 Å, U_{iso}(H) = 1.5 U_{eq}(C) (for methyl). The methyl groups were allowed to rotate about their local threefold axis. Details of the crystallographic data collection, structural determination, and refinement for 1 and 2 are given in Table 2, whereas selected bond lengths and angles for them are listed in Tables S1 and S2, Supplementary Materials.

Powder X-ray diffraction (XRPD) measurements on 1 and 2 were performed on a PANalytical Empyrean X-ray diffractometer by using Cu−K$_\alpha$ radiation (λ = 1.5418 Å), in which the X-ray tube was operated at 40 kV and 30 mA ranging from 5 to 50° (Figures 2 and S9).

3.4. Physical Measurements

IR spectra were recorded on a Nicolet iS5 FT–IR spectrophotometer in the spectral range 4000–400 cm^{-1} with the samples in the form of KBr pellets (Figure S1, Supplementary Materials). HRMS analysis (Figure S2, Supplementary Materials) was performed on a Waters Xevo G2 Q-TOF mass spectrometer (Waters Corporation, Milford, MA, USA) with an ESI ion source. Full-scan MS data were collected from 100 to 1000 Da in positive ion mode with a scan time of 0.5 s. To ensure accurate mass measurements, data were collected in centroid mode and mass was corrected during acquisition using a leucine enkephalin solution as an external reference (Lock-Spray^{TM}), with the reference ion at m/z 556.2771 Da ([M + H]$^+$). Elemental analyses (C, H, N) were performed on a Perkin–Elemer CHN–2400 analyzer. The electronic spectra were obtained using Nicolet Evolution 220 in the range 240–1000 nm in DMSO (Figure S10, Supplementary Materials).

Table 2. Crystal data and structure refinement for **1–2**.

	1	2
Empirical formula	$C_{22}H_{20}N_2O_3V$	$C_{22}H_{20}N_2O_3V$
Formula weight	411.34	411.34
T, K	295.0(2)	295.0(2)
Wavelength, Å	0.71073	0.71073
Crystal system	Monoclinic	Monoclinic
Space group	$C2/c$	$P2_1/c$
Unit cell dimensions, Å and °		
a	18.8566(15)	15.6046(11)
b	8.2237(4)	8.1475(5)
c	13.2105(8)	16.5949(16)
β	113.906(5)	117.340(11)
V, Å^3	1872.8(2)	1874.2(3)
Z	4	4
D_c, g cm^{-3}	1.459	1.458
Absorption coefficient, mm^{-1}	0.556	0.555
$F(000)$	852	852
Crystal size, mm	0.283 × 0.137 × 0.055	0.162 × 0.089 × 0.079
θ range for data collection °	3.46 to 25.05	3.49 to 25.05
Index ranges	$-22 \leq h \leq 22$ $-9 \leq k \leq 9$ $-15 \leq l \leq 15$	$-18 \leq h \leq 17$ $-8 \leq k \leq 9$ $-19 \leq l \leq 18$
Reflections collected	6154	7065
Independent reflections	1653 [R_{int} = 0.0221]	3291 [R_{int} = 0.0479]
Completeness to 2θ	99.7	99.3
Min. and max. transm.	0.712 and 1.000	0.588 and 1.000
Data/restraints/parameters	1653/0/130	3291/0/257
Goodness-of-fit on F^2	1.086	1.002
Final R indices [$I > 2\sigma(I)$]		
R1	0.0326	0.0489
wR2	0.0931	0.1065
R indices (all data)		
R1	0.0355	0.0834
wR2	0.0950	0.1202
Largest diff. peak and hole, e Å^{-3}	0.50 and −0.24	0.34 and −0.32
CCDC number	1971585	1971586

3.5. EPR Spectroscopy

Electron paramagnetic resonance (EPR) spectra of the oxidovanadium(IV) complexes were measured using a Bruker ELEXYS E 500 operating at the X-band frequency (9.7 GHz). The solid compounds **1** and **2** dissolved in water and a few drops of DMSO, were added to the samples to ensure good glass formation at liquid nitrogen temperature [101]. A microwave frequency of 6.231 GHz, power of 10 mW and modulation amplitude of 8 G was used. Anisotropic spectra were recorded on frozen solutions at 77 K using quartz Dewar and glass capillary tubes at room temperature. An analysis of the EPR spectra was carried out using the WinEPR SimFonia software package, version 1.26b [102].

3.6. Biological Assays

3.6.1. Cell Culture

The human colorectal carcinoma derived cancer cell line (HCT116) and human normal dermal fibroblasts (Ref. PCS-201-010) were purchased from American Type Culture Collection (ATCC, Manassas, VA, USA) and grown in Dulbecco's modified Eagle medium (DMEM). The human ovarian carcinoma derived cancer cell line (A2780) was purchased from Merck (Darmstadt, Germany) and cultivated in Roswell Park Memorial Institute (RPMI) 1640 culture medium. All media were supplemented with 10% fetal bovine serum and a 1% Pen/Strep solution (all media and supplements were from Thermo Fischer Scientific, Waltham, MA, USA). Cell cultures were maintained at 37 °C, in a humified atmosphere of 5% (v/v) CO_2 [103,104].

3.6.2. Viability Assays

Normal dermal fibroblasts and cancer cell lines HCT116 and A2780 were seeded in 96-well plates with a density of 7500 cells per well. After 24 h, culture media was replaced, and cells were exposed to different concentrations of complexes **1–3** or DMSO 0.1% (v/v) (vehicle control) or cisplatin (positive control) for 48 h (Figure S14). Cells exposed to cisplatin were used as a positive control [103,104]. After 48 h of incubation, the CellTiter 96® Aqueous Non-Radioactive Proliferation assay (Promega, Madison, WI, USA) was used to determine cellular viability through the production of formazan through the reduction of 3-(4,5-dimethylthiazol-2-yl)-5-(3-carboxymethoxyphenyl)-2-(4-sulfophenyl)-2H-tetrazolium, inner salt (MTS) by dehydrogenases present in metabolically active cells [103,104]. The amount of formazan can be determined by its absorbance at 492 nm in an Infinite M200 microplate reader (Tecan, Mannedorf, Switzerland) [103,104]. The biological activity of the complexes was compared using the half maximal inhibitory concentration of cellular proliferation (IC_{50}) determined with Prism 8.2.1 software for windows (GraphPad Software, La Jolla, CA, USA).

3.6.3. Vanadium Detection in the HCT116 Cell Line by ICP-AES

The internalization of the vanadium complexes in HCT116 cells was evaluated by an inductive plasma atomic emission spectrometry technique (ICP-AES). HCT116 cells were seeded in 25 cm^2 culture flasks with a cellular density of 1×10^5 cells per mL. After 24 h of incubation at 37 °C in a humified atmosphere of 5% (v/v) CO_2, the culture medium was replaced with fresh culture medium containing a $20 \times IC_{50}$ at 48 h concentration of the complexes **1–3** or 0.1% (v/v) of DMSO (vehicle control) followed by incubations of 3 h or 6 h at 37 or 4 °C. We used this concentration to ensure that we were above the detection limit of the technique. This concentration ($20 \times IC_{50}$) does not induce a loss of cell viability for the selected time points. The culture medium was then collected, and cells detached with trypsin and centrifuged at $700 \times g$ for 5 min at 15 °C. The supernatant (culture medium plus trypsin) and the pellet (cells) were stored separately at −20 °C until freshly made *aqua regia* was added to all the samples as previously described [104]. The vanadium quantification in each fraction was performed by ICP-AES, as a paid service.

3.6.4. Evaluation of Apoptosis Induction in the HCT116 Cell Line by Flow Cytometry

Apoptosis induction in HCT116 cells was evaluated using the Annexin V-Alexa fluor 488/PI dead cell apoptosis assay (Thermo Fischer Scientific). Briefly, HCT116 cells were seeded in 6-well plates (2×10^5 cells per well) and then incubated 48 h with the IC_{50} concentrations of complexes **1–3** at 37 °C in a humified atmosphere of 5% (v/v) CO_2. Cells were also incubated with DMSO 0.1% (vehicle control) and cisplatin 15 µM (positive control). Following the manufacturer's instructions, after this incubation period, cells were detached with trypsin, washed with PBS 1× and incubated 15 min at room temperature with Annexin V-Alexa fluor 488 assay solution and 10 µg mL^{-1} propidium iodide [97,104]. An Attune acoustic focusing cytometer (ThermoFisher Scientific, Waltham, MA, USA) was used to analyze cells and the resulting information was processed with the respective Attune Cytometric Software 2.1 (ThermoFisher Scientific, Waltham, MA, USA).

3.6.5. Autophagy Induction Evaluation in the HCT116 Cell Line by Flow Cytometry

Autophagy induction in HCT116 cells was evaluated using the Autophagy Assay kit (Abcam), according to the manufacturer's instructions. HCT116 cells were seeded in 6-well plates at a cellular density of 2×10^5 cells per well. After 24 h, cells were incubated with the IC_{50} concentrations of complexes **1–3** for 48 h at 37 °C in a humified atmosphere of 5% (v/v) CO_2. In addition to the DMSO 0.1% v/v vehicle control, cisplatin 15 µM and rapamycin 0.5 µM were performed as positive controls. After 48 h of incubation, cells were detached from the wells with trypsin and washed with Assay Buffer 1× before being incubated 30 min at 37 °C in DMEM medium with the Green Stain solution. Cells were then washed and resuspended in Assay Buffer 1× and this was followed by analysis in an Attune acoustic focusing cytometer (ThermoFisher Scientific), and the results were analyzed with the respective instrument software (Attune Cytometric Software, version 2.1).

3.6.6. Intracellular Reactive Oxygen Species (ROS) Production Evaluation in the HCT116 Cell Line by Flow Cytometry

The induction of ROS in HCT116 cells was evaluated indirectly by flow cytometry using a specific dye, 2′,7′-dichlorodihydrofluorescein diacetate (H_2DCF-DA) (Thermo Fischer Scientific). In the presence of ROS, intracellular esterases remove the acetate groups of the dye which leads to an increased fluorescence [97,104]. HCT116 cells were seeded in 6-well plates (4×10^5 cells per well) for the initial 24 h of incubation. Cells were incubated with DMSO 0.1% (v/v) (vehicle control), 50 µM H_2O_2 and 15 µM cisplatin (positives controls) and the complexes **1–3** at their IC_{50} concentrations for 48 h at 37 °C in a humified atmosphere of 5% (v/v) CO_2. Cells were then detached from the wells with trypsin and washed with PBS 1× before incubation with 100 µM of H_2DCF-DA for 20 min at 37 °C, then processed in an Attune acoustic focusing cytometer (ThermoFisher Scientific), with the resulting data analyzed using the respective software (Attune Cytometric Software, version 2.1).

3.6.7. Mitochondrial Membrane Potential Evaluation in the HCT116 Cell Line by Flow Cytometry

The mitochondrial membrane potential in HCT116 cells was evaluated using 5,5′,6,6′-Tetrachloro-1,1′,3,3′-tetraethylbenzimidazolocarbocyanine iodide, JC-1 (Abnova Corporation, Walnut, CA, USA). The mitochondrial potential is an important parameter of mitochondrial function, and is thus used as an indicator of cell health. In healthy cells with high potential, JC-1 shows aggregates with intense red fluorescence. On the other hand, in cells with low mitochondrial membrane potential (for example cells in apoptosis), JC-1 remains in the monomeric form, which exhibits green fluorescence[104] [105]. Briefly, HCT116 cells were seeded in 6-well plates at a cellular density of 2×10^5 cells per well and then incubated 48 h with the IC_{50} concentrations of vanadium complexes **1–3** at 37 °C in a humified atmosphere of 5% (v/v) CO_2, having as controls DMSO 0.1% (vehicle control), cisplatin 15 µM and doxorubicin 0.4 µM (positive controls). Afterwards cells were detached with trypsin, washed with PBS 1× and incubated 20 min at 37 °C in DMEM medium with the JC-1 solution. Cells were again washed and resuspend in DMEM medium without phenol red and analyzed in an Attune acoustic focusing cytometer (ThermoFisher Scientific).

3.7. Catalytic Studies

The total volume of the reaction solution was 5 mL (Caution: the combination of air or molecular oxygen and H_2O_2 with organic compounds at elevated temperatures may be explosive!). Cylindrical glass vessels with vigorous stirring of the reaction mixture were used for the oxidation of alkanes with hydrogen peroxide which were typically carried out in air in a thermostated solution. Initially, a portion of 50% aqueous solution of hydrogen peroxide was added to the solution of the catalyst and substrate in acetonitrile. The aliquots of the reaction solution were analyzed by GC (a 3700 instrument, fused silica capillary column FFAP/OV-101 20/80 w/w, 30 m × 0.2 mm × 0.3 µm; argon as a carrier gas. Attribution of peaks was made by comparison with chromatograms of authentic samples). Usually samples were analyzed twice, i.e., before and after the addition portion by portion of the excess of solid PPh$_3$. This method was proposed and used by one of us previously [106,107].

Alkyl hydroperoxides are transformed in the GC injector into a mixture of the corresponding ketone and alcohol. Due to this, we quantitatively reduced the reaction samples with PPh$_3$ to obtain the corresponding alcohol. This method allows us to calculate the real concentrations not only of the hydroperoxide, but of the alcohols and ketones present in the solution at a given moment. An example is shown in Figure 16.

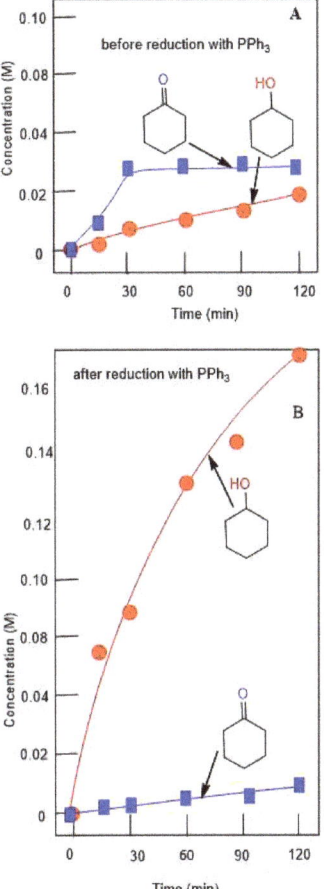

Figure 16. Accumulation of cyclohexanol and cyclohexanone in the oxidation of cyclohexane (0.46 M) with H$_2$O$_2$ (2.0 M) catalyzed by complex **2** (5 × 10^{-4} M) at 50 °C in acetonitrile. Concentrations of products were measured by GC before (Graph (**A**)) and after (Graph (**B**)) the reduction of the reaction samples with solid PPh$_3$.

4. Conclusions

The studies revealed that methyl-functionalization of 8-hydroxyquinoline facilitates the formation of five-coordinated oxidovanadium(IV) complexes in the reaction of bis(acety lacetonato)oxidovanadium(IV) 2,6-(Me)$_2$-quin, 2,5-(Me)$_2$-quin and 2-Me-quin.

Complexes **1–3** catalyze very efficient transformation of saturated hydrocarbons into alkyl hydroperoxides, alcohols and ketones. The reaction requires addition of PCA and occurs with the participation of hydroxyl radicals.

Here we show that complex **3** effectively reduced the viability of a HCT116 colorectal cancer cell line with low or no cytotoxicity in normal dermal fibroblasts or in an ovarian carcinoma cell line, respectively. On the other hand, complexes **1** and **2** bearing additional methyl groups show lower antiproliferative activities in HCT116 cells, and complex **2** shows some degree of cytotoxicity towards primary fibroblasts. Complexes internationalization is probably associated with passive transport, with more than 90% of complexes **2** and **3** being in the cellular fraction after a 6 h incubation. All complexes can increase in the production of intracellular ROS which can trigger apoptosis and autophagy in the HCT116 cell line. Mitochondrial membrane potential was not significantly altered by the three complexes in HCT116 cells which may indicate that the intrinsic pathway is not activated.

Supplementary Materials: The following are available online. Figure S1: IR spectra of complexes **1–3** and free ligands, Figure S2: HRMS of complexes **1–3**, Tables S1–S3: Selected bond lengths and angles for **1–3**, Table S4: Short intra- and intermolecular hydrogen bonds, Table S5: Short π•••π interactions, Table S6: X—Y•••Cg(J)(π-ring) interactions, Figures S3–S5: Hirshfeld surface and 2D fingerprint plot for **1–3**, together with the relative contributions of various intermolecular interactions with the Hirshfeld surface, Figures S6–S8: View of the intermolecular interactions and packing in **1–3**, Table S7: The selected structural parameters of five-coordinated vanadium(IV) complexes, Figure S9: X-ray powder diffraction patterns of **1–3**, Figure S10: EPR frozen solution spectra (at 77 K) in acetonitrile, Figure S11: UV–Vis spectra of complexes **1–3**, Figure S12: UV–Vis stability spectra in DMSO of **1–3**, Figure S13: Antiproliferative effect of complexes **1–3** in normal dermal fibroblasts, after 48 h, evaluated by the MTS method, Figure S14: Cell viability of HCT116 cells.

Author Contributions: Synthesis of V(IV) complexes, J.P.-G., K.C. and B.M.; general characterization, X-ray analysis and UV–Vis studies, J.P.-G., K.C. and B.M.; synthesis of 2,5-dimethyl-8-hydroxyquinoline and 2,6-dimethyl-8-hydroxyquinoline, J.E.N.; HRMS analysis, K.E.; EPR studies, A.B. and A.L.(Agnieszka Lewińska); studies of catalytic oxidations, L.S.S., N.S.I. and G.B.S.; biological data assessment, A.L.(André Luz) and L.R.R.; interpretation and writing of initial and final version of the manuscript, A.R.F. and P.V.B. All authors have read and agreed to the published version of the manuscript.

Funding: This research was funded by the Russian Foundation for Basic Research (Grant Nos. 19-03-00142), the Ministry of Education and Science of the Russian Federation (project code RFMEFI61917X0007), as well as by the Initiative Program in the frames of the State Task 0082-2014-0007, "Fundamental regularities of heterogeneous and homogeneous catalysis" and also by the Program of Fundamental Research of the Russian Academy of Sciences for 2013–2020 on the research issue of IChP RAS No. 47.16. State registration number of the Center of Information Technologies and Systems for Executive Power Authorities (CITIS): AAAA-A17-117040610283-3; GC analysis was performed with the financial support from the Ministry of Education and Science of the Russian Federation using the equipment of the Center for molecular composition studies of the A.N. Nesmeyanov INEOS RAS. This work is also financed by national funds from FCT-Fundação para a Ciência e a Technological, I.P., in the scope of the project UIDP/04378/2020 and UIDB/04378/2020 of the Research Unit on Applied Molecular Biosciences-UCIBIO and the project LA/P/0140/2020 of the Associate Laboratory Institute for Health and Bioeconomy-i4HB.

Institutional Review Board Statement: Not applicable.

Informed Consent Statement: Not applicable.

Data Availability Statement: Crystallographic data for **1** and **2** have been deposited with the Cambridge Crystallographic Data Center, CCDC 1971585–1971586. Copies of this information may be obtained free of charge from the Director, CCDC, 12 Union Road, Cambridge CB2 1EZ, UK (Fax: +44-1223-336033; e-mail: deposit@ccdc.cam.ac.uk or www.ccdc.cam.ac.uk, 12 December 2019).

Acknowledgments: Authors are thankful for support by RFBR according to Research Project Grant No. 19-03-00142; the Ministry of Education and Science of the Russian Federation (project code RFMEFI61917X0007), as well as by the Initiative Program in the frames of the State Tasks 0082-2014-0004 and 0082-2014-0007 "Fundamental regularities of heterogeneous and homogeneous catalysis"; the Program of Fundamental Research of the Russian Academy of Sciences for 2013–2020 on the research issue of IChP RAS No. 47.16.; and the A.N. Nesmeyanov Institute RAS (Institute of Organoelement Compounds, Russian Academy of Sciences).

Conflicts of Interest: The authors declare no conflict of interest.

Sample Availability: Samples of the compounds **1–3** are available from the authors.

References

1. Butler, A.; Carrano, C. Coordination chemistry of vanadium in biological systems. *Coord. Chem. Rev.* **1991**, *109*, 61–105. [CrossRef]
2. Chatterjee, M.; Ghosh, S.; Wu, B.-M.; Mak, T.C.W. A structural and electrochemical study of some oxovanadium (IV) heterochelate complexes. *Polyhedron* **1998**, *17*, 1369–1374. [CrossRef]
3. Liu, S.-X.; Gao, S. Synthesis, crystal structure and spectral properties of VO (acetylacetone benzoylhydrazone) (8-quinolinol). *Polyhedron* **1998**, *17*, 81–84. [CrossRef]
4. Bhattacharyya, S.; Mukhopadhyay, S.; Samanta, S.; Weakley, T.J.R.; Chaudhury, M. Synthesis, characterization, and reactivity of mononuclear O, N-chelated vanadium(IV) and -(III) complexes of methyl 2-aminocyclopent-1-ene-1-dithiocarboxylate based ligand: Reporting an example of conformational isomerism in the solid state. *Inorg. Chem.* **2002**, *41*, 2433–2440. [CrossRef] [PubMed]
5. Horn, A., Jr.; Filgueiras, C.A.L.; Wardell, J.L.; Herbst, M.H.; Vugman, N.V.; Santos, P.S.; Lopes, J.G.S.; Howie, R.A. A fresh look into VO (salen) chemistry: Synthesis, spectroscopy, electrochemistry and crystal structure of [VO(salen)(H$_2$O)]Br·0.5CH$_3$CN. *Inorg. Chim. Acta* **2004**, *357*, 4240–4246. [CrossRef]
6. Elias, H.; Schwartze-Eidam, S.; Wannowius, K.J. Kinetics and mechanism of ligand substitution in bis(N-alkylsalicylaldiminato)ox ovanadium(IV) complexes. *Inorg. Chem.* **2003**, *42*, 2878–2885. [CrossRef] [PubMed]
7. Correia, I.; Pessoa, J.C.; Duarte, M.T.; Henriques, R.T.; Fátima, M.; Piedade, M.; Veiros, L.F.; Jakusch, T.; Kiss, T.; Dörnyei, Á.; et al. N,N'-ethylenebis(pyridoxylideneiminato) and N,N'-ethylenebis(pyridoxylaminato): Synthesis, characterization, potentiometric, spectroscopic, and DFT studies of their vanadium(IV) and vanadium(V) complexes. *Chem. Eur. J.* **2004**, *10*, 2301–2317. [CrossRef]
8. Yucesan, G.; Armatas, N.G.; Zubieta, J. Hydrothermal synthesis of molecular oxovanadium compounds. The crystal and molecular structures of [VO$_2$(terpy)]NO$_3$, [VO(terpy)(OH$_3$PC$_6$H$_5$)$_2$], [{Cu(H$_2$O)(terpy)}V$_2$O$_6$], [{Cu(ttbterpy)}V$_2$O$_6$] and [{Cu(ttbterpy)}VO$_2$(HO$_3$PCH$_2$PO$_3$)]·H$_2$O (terpy = 2,2':6',2''-terpyridine; ttbterpy = 4,4',4''-tri-*tert*-butyl-2,2':6',2''-terpyridine). *Inorg. Chim. Acta* **2006**, *359*, 4557–4564.
9. Ghosh, T.; Mondal, B.; Ghosh, T.; Sutradhar, M.; Mukherjee, G.; Drew, M.G.B. Synthesis, structure, solution chemistry and the electronic effect of *para* substituents on the vanadium center in a family of mixed-ligand [V^{V}O(ONO)(ON)] complexes. *Inorg. Chim. Acta* **2007**, *360*, 1753–1761. [CrossRef]
10. Rubčić, M.; Milić, D.; Horvat, G.; Đilović, I.; Galić, N.; Tomšić, V.; Cindrić, M. Vanadium-induced formation of thiadiazole and thiazoline compounds. Mononuclear and dinuclear oxovanadium(V) complexes with open-chain and cyclized thiosemicarbazone ligands. *Dalton Trans.* **2009**, *2009*, 9914–9923. [CrossRef]
11. Hakimi, M.; Kukovec, B.-M.; Rezvaninezhad, M.; Schuh, E.; Mohr, F. Preparation, structural and spectroscopic characterization of vanadium(IV) and vanadium(V) complexes with dipicolinic acid. *Z. Anorg. Allg. Chem.* **2011**, *637*, 2157–2162. [CrossRef]
12. Sanna, D.; Buglyó, P.; Tomaz, A.I.; Costa Pessoa, J.; Borović, S.; Micerae, G.; Garribba, E. V^{IV}O and Cu^{II} complexation by ligands based on pyridine nitrogen donors. *Dalton Trans.* **2012**, *41*, 12824–12838. [CrossRef] [PubMed]
13. Tutusaus, O.; Ni, C.; Szymczak, N.K. A Transition metal Lewis acid/base triad system for cooperative substrate binding. *J. Am. Chem. Soc.* **2013**, *135*, 3403–3406. [CrossRef]
14. Bhattacharya, K.; Maity, M.; Abtab, S.M.T.; Majee, M.C.; Chaudhury, M. Homo- and heterometal complexes of oxido–metal ions with a triangular [V(V)O–MO–V(V)O] [M = V(IV) and Re(V)] core: Reporting mixed-oxidation oxido–vanadium(V/IV/V) compounds with valence trapped structures. *Inorg. Chem.* **2013**, *52*, 9597–9605. [CrossRef]
15. Sheng, G.-H.; Cheng, X.-S.; You, Z.-L.; Zhu, H.-L. Two isomeric structures of oxovanadium(V) complexes with hydrazone and 8-hydroxyquinoline ligands. *J. Struct. Chem.* **2015**, *56*, 942–947. [CrossRef]
16. Sheppard, B.J.H.; Shaver, M.P.; Pearson, J.K. Assessment and application of density functional theory for the prediction of structure and reactivity of vanadium complexes. *J. Phys. Chem. A* **2015**, *119*, 8537–8546. [CrossRef] [PubMed]
17. Sutradhar, M.; Pombeiro, A.J.L. Coordination chemistry of non-oxido, oxido and dioxidovanadium(IV/V) complexes with azine fragment ligands. *Coord. Chem. Rev.* **2014**, *265*, 89–124. [CrossRef]

18. Kolesa-Dobravc, T.; Lodyga-Chruscinska, E.; Symonowicz, M.; Sanna, D.; Meden, A.; Perdih, F.; Garribba, E. Synthesis and characterization of V(IV)O complexes of picolinate and pyrazine derivatives. Behavior in the solid state and aqueous solution and biotransformation in the presence of blood plasma proteins. *Inorg. Chem.* **2014**, *53*, 7960–7976. [CrossRef]
19. Chang, Y.-P.; Furness, L.; Levason, W.; Reid, G.; Zhang, W. Complexes of vanadium(IV) oxide difluoride with neutral N- and O-donor ligands. *J. Fluor. Chem.* **2016**, *191*, 149–160. [CrossRef]
20. Dash, S.P.; Majumder, S.; Banerjee, A.; Fernanda, M.; Carvalho, N.N.; Adaão, P.; Costa Pessoa, J.; Brzezinski, K.; Garribba, E.; Reuter, H.; et al. Chemistry of monomeric and dinuclear non-oxido vanadium(IV) and oxidovanadium(V) aroylazine complexes: Exploring solution behavior. *Inorg. Chem.* **2016**, *55*, 1165–1182. [CrossRef]
21. Süss-Fink, G.; Gonzalez Cuervo, L.; Therrien, B.; Stoeckli-Evans, H.; Shul'pin, G.B. Mono and oligonuclear vanadium complexes as catalysts for alkane oxidation: Synthesis, molecular structure, and catalytic potential. *Inorg. Chim. Acta* **2004**, *357*, 475–484. [CrossRef]
22. Crans, D.C.; Smee, J.J.; Gaidamauskas, E.; Yang, L. The chemistry and biochemistry of vanadium and the biological activities exerted by vanadium compounds. *Chem. Rev.* **2004**, *104*, 849–902. [CrossRef] [PubMed]
23. Hanson, S.K.; Baker, R.T.; Gordon, J.C.; Scott, B.L.; Silks, L.A.; Thorn, D.L. Mechanism of alcohol oxidation by dipicolinate vanadium (V): Unexpected role of pyridine. *J. Am. Chem. Soc.* **2010**, *132*, 17804–17816. [CrossRef]
24. Zhang, G.; Scott, B.L.; Wu, R.; Silks, L.A.; Hanson, S.K. Aerobic oxidation reactions catalyzed by vanadium complexes of bis (phenolate) ligands. *Inorg. Chem.* **2012**, *51*, 7354–7361. [CrossRef]
25. Grivani, G.; Ghavami, A.; Kučeráková, M.; Dušek, M.; Dehno Khalaji, A. Synthesis, characterization, crystal structure determination, thermal study and catalytic activity of a new oxidovanadium Schiff base complex. *J. Mol. Struct.* **2014**, *1076*, 326–332. [CrossRef]
26. Tuskaev, V.A.; Kolosov, N.A.; Kurmaev, D.A.; Gagieva, S.C.; Khrustalev, V.N.; Ikonnikov, N.S.; Efimov, N.N.; Ugolkova, E.A.; Minin, V.V.; Bulychev, B.M. Vanadium (IV), (V) coordination compounds with 8-hydroxyquinoline derivative: Synthesis, structure and catalytic activity in the polymerization of ethylene. *J. Organomet. Chem.* **2015**, *798*, 393–400. [CrossRef]
27. Mandal, M.; Nagaraju, V.; Karunakar, G.V.; Sarma, B.; Borah, B.J.; Bania, K.K. Electronic, conjugation, and confinement effects on structure, redox, and catalytic behavior of oxido-vanadium (IV) and-(V) chiral Schiff base complexes. *J. Phys. Chem. C* **2015**, *119*, 28854–28870. [CrossRef]
28. Sutradhar, M.; Martins, L.M.; Guedes da Silva, M.F.C.; Pombeiro, A.J.L. Oxidovanadium complexes with tridentate aroylhydrazone as catalyst precursors for solvent-free microwave-assisted oxidation of alcohols. *Appl. Catal. A Gen.* **2015**, *493*, 50–57. [CrossRef]
29. Chieregato, A.; Lopez Nieto, J.M.; Cavani, F. Mixed-oxide catalysts with vanadium as the key element for gas-phase reactions. *Coord. Chem. Rev.* **2015**, *301–302*, 3–23. [CrossRef]
30. Elkurtehi, A.I.; Walsh, A.G.; Dawe, L.N.; Kerton, F.M. Vanadium Aminophenolate complexes and their catalytic activity in aerobic and H_2O_2-mediated oxidation reactions. *Eur. J. Inorg. Chem.* **2016**, *2016*, 3123–3130. [CrossRef]
31. Wang, Y.; Lin, X.-M.; Bai, F.-Y.; Sun, L.-X. Novel vanadium complexes with rigid carboxylate ligands: Synthesis, structure and catalytic bromine dynamics of phenol red. *J. Mol. Struct.* **2017**, *1149*, 379–386. [CrossRef]
32. Schmidt, A.-C.; Hermsen, M.; Rominger, F.; Dehn, R.; Teles, J.H.; Schaäfer, A.; Trapp, O.; Schaub, T. Synthesis of mono-and dinuclear vanadium complexes and their reactivity toward dehydroperoxidation of alkyl hydroperoxides. *Inorg. Chem.* **2017**, *56*, 1319–1332. [CrossRef]
33. Pessoa, J.C. Thirty years through vanadium chemistry. *J. Inorg. Biochem.* **2015**, *147*, 4–24. [CrossRef] [PubMed]
34. Rehder, D. The coordination chemistry of vanadium as related to its biological functions. *Coord. Chem. Rev.* **1999**, *182*, 297–322. [CrossRef]
35. Xie, M.-J.; Niu, Y.-F.; Yang, X.-D.; Liu, W.-P.; Li, L.; Gao, L.-H.; Yan, S.-P.; Meng, Z.-H. Effect of the chloro-substitution on lowering diabetic hyperglycemia of vanadium complexes with their permeability and cytotoxicity. *Eur. J. Med. Chem.* **2010**, *45*, 6077–6084. [CrossRef]
36. Banik, B.; Sasmal, P.K.; Roy, S.; Majumdar, R.; Dighe, R.R.; Chakravarty, A.R. Terpyridine oxovanadium (IV) complexes of phenanthroline bases for cellular imaging and photocytotoxicity in HeLa cells. *Eur. J. Inorg. Chem.* **2011**, *2011*, 1425–1435. [CrossRef]
37. Correia, I.; Adão, P.; Roy, S.; Wahba, M.; Matos, C.; Maurya, M.R.; Marques, F.; Pavan, F.R.; Leite, C.Q.F.; Avecilla, F.; et al. Hydroxyquinoline derived vanadium(IV and V) and copper(II) complexes as potential anti-tuberculosis and anti-tumor agents. *J. Inorg. Biochem.* **2014**, *141*, 83–93. [CrossRef]
38. Fik, M.A.; Gorczyński, A.; Kubicki, M.; Hnatejko, Z.; Wadas, A.; Kulesza, P.J.; Lewińska, A.; Giel-Pietraszuk, M.; Wyszko, E.; Patroniak, V. New vanadium complexes with 6,6''-dimethyl-2,2':6',2''-terpyridine in terms of structure and biological properties. *Polyhedron* **2015**, *97*, 83–93. [CrossRef]
39. Kioseoglou, E.; Petanidis, S.; Gabriel, C.; Salifoglou, A. The chemistry and biology of vanadium compounds in cancer therapeutics. *Coord. Chem. Rev.* **2015**, *301–302*, 87–105. [CrossRef]
40. Reytman, L.; Braitbard, O.; Hochman, J.; Tshuva, E.Y. Highly effective and hydrolytically stable vanadium (V) amino phenolato antitumor agents. *Inorg. Chem.* **2016**, *55*, 610–618. [CrossRef]
41. Kumar, A.; Pant, I.; Dixit, A.; Banerjee, S.; Banik, B.; Saha, R.; Kondaiah, P.; Chakravarty, A.R. Terpyridyl oxovanadium (IV) complexes for DNA crosslinking and mito-targeted photocytotoxicity. *J. Inorg. Biochem.* **2017**, *174*, 45–54. [CrossRef] [PubMed]

42. Hong, X.-L.; Liu, L.-J.; Lu, W.-G.; Wang, X.-B. A vanadium (V) terpyridine complex: Synthesis, characterization, cytotoxicity in vitro and induction of apoptosis in cancer cells. *Transit. Met. Chem.* **2017**, *42*, 459–467. [CrossRef]
43. Ni, L.; Zhao, H.; Tao, L.; Li, X.; Zhou, Z.; Sun, Y.; Chen, C.; Wei, D.; Liu, Y.; Diao, G. Synthesis, in vitro cytotoxicity, and structure–activity relationships (SAR) of multidentate oxidovanadium(IV) complexes as anticancer agents. *Dalton Trans.* **2018**, *47*, 10035–10045. [CrossRef]
44. El-Deen, I.M.; Shoair, A.F.; El-Bindary, M.A. Synthesis, characterization and biological properties of oxovanadium(IV) complexes. *J. Mol. Struct.* **2019**, *1180*, 420–437. [CrossRef]
45. Floris, B.; Sabuzi, F.; Coletti, A.; Conte, V. Sustainable vanadium-catalyzed oxidation of organic substrates with H_2O_2. *Catal. Today* **2017**, *285*, 49–56. [CrossRef]
46. Hasnaoui, A.; Idouhli, R.; Nayad, A.; Ouahine, H.; Khadiri, M.-E.; Abouelfidab, A.; Elfirdoussi, L.; Ait Ali, M. Di-nuclear water-soluble oxovanadium (V) Schiff base complexes: Electrochemical properties and catalytic oxidation. *Inorg. Chem. Commun.* **2020**, *119*, 108134.
47. Crans, D. Chemistry and insulin-like properties of vanadium (IV) and vanadium (V) compounds. *J. Inorg. Biochem.* **2000**, *80*, 123–131. [CrossRef]
48. Banik, B.; Somyajit, K.; Koleyc, D.; Nagaraju, G.; Chakravarty, A.R. Cellular uptake and remarkable photocytotoxicity of pyrenylter pyridine oxovanadium(IV) complexes of dipyridophenazine bases. *Inorg. Chim. Acta* **2012**, *393*, 284–293. [CrossRef]
49. Banik, B.; Somyajit, K.; Nagaraju, G.; Chakravarty, A.R. Oxovanadium (IV) catecholates of terpyridine bases for cellular imaging and photocytotoxicity in red light. *RSC Adv.* **2014**, *4*, 40120–40131. [CrossRef]
50. Banik, B.; Somyajit, K.; Nagaraju, G.; Chakravarty, A.R. Oxovanadium (IV) complexes of curcumin for cellular imaging and mitochondria targeted photocytotoxicity. *Dalton Trans.* **2014**, *43*, 13358–13369. [CrossRef]
51. Banik, B.; Somyajit, K.; Hussain, A.; Nagaraju, G.; Chakravarty, A.R. Carbohydrate-appended photocytotoxic (imidazophenanthroline)-oxovanadium (IV) complexes for cellular targeting and imaging. *Dalton Trans.* **2014**, *43*, 1321–1331. [CrossRef]
52. Balaji, B.; Balakrishnan, B.; Perumalla, S.; Karande, A.A.; Chakravarty, A.R. Photoactivated cytotoxicity of ferrocenyl-terpyridine oxovanadium(IV) complexes of curcuminoids. *Eur. J. Med. Chem.* **2014**, *85*, 458–467. [CrossRef]
53. Balaji, B.; Balakrishnan, B.; Perumalla, S.; Karande, A.A.; Chakravarty, A.R. Photocytotoxic oxovanadium(IV) complexes of ferrocenyl-terpyridine and acetylacetonate derivatives. *Eur. J. Med. Chem.* **2015**, *92*, 332–341. [CrossRef] [PubMed]
54. Levina, A.; McLeod, A.I.; Gasparini, S.J.; Nguyen, A.; Manori De Silva, W.G.; Aitken, J.B.; Harris, H.H.; Glover, C.; Johannessen, B.; Lay, P.A. Reactivity and speciation of anti-diabetic vanadium complexes in whole blood and its components: The important role of red blood cells. *Inorg. Chem.* **2015**, *54*, 7753–7766. [CrossRef]
55. Pessoa, J.C.; Etcheverry, S.; Gambino, D. Vanadium compounds in medicine. *Coord. Chem. Rev.* **2015**, *301–302*, 24–48. [CrossRef] [PubMed]
56. Thompson, K.H.; Lichter, J.; LeBel, C.; Scaife, M.C.; McNeill, J.H.; Orvig, C. Vanadium treatment of type 2 diabetes: A view to the future. *J. Inorg. Biochem.* **2009**, *103*, 554–558. [CrossRef] [PubMed]
57. Fukui, K.; Fujisawa, Y.; Ohya-Nishiguchi, H.; Kamada, H.; Sakurai, H. In vivo coordination structural changes of a potent insulin-mimetic agent, bis(picolinato)oxovanadium(IV), studied by electron spin-echo envelope modulation spectroscopy. *J. Inorg. Biochem.* **1999**, *77*, 215–224. [CrossRef]
58. Thompson, K.H.; McNeill, J.H.; Orvig, C. Vanadium compounds as insulin mimics. *Chem. Rev.* **1999**, *99*, 2561–2571. [CrossRef] [PubMed]
59. Thompson, K.H.; Orvig, C. Coordination chemistry of vanadium in metallopharmaceutical candidate compounds. *Coord. Chem. Rev.* **2001**, *219–221*, 1033–1053. [CrossRef]
60. Sakurai, H.; Kojima, M.; Yoshikawa, Y.; Kawabe, K.; Yasui, H. Antidiabetic vanadium (IV) and zinc (II) complexes. *Coord. Chem. Rev.* **2002**, *226*, 187–198. [CrossRef]
61. Kiss, T.; Jakusch, T.; Hollender, D.; Enyedy, EÉA.; Horvath, L. Comparative studies on the biospeciation of antidiabetic VO(IV) and Zn(II) complexes. *J. Inorg. Biochem.* **2009**, *103*, 527–535. [CrossRef] [PubMed]
62. Choroba, K.; Raposo, L.R.; Palion-Gazda, J.; Malicka, E.; Erfurt, K.; Machura, B.; Fernandes, A.R. In vitro antiproliferative effect of vanadium complexes bearing 8-hydroxyquinoline-based ligands–the substituent effect. *Dalton Trans.* **2020**, *49*, 6596–6606. [CrossRef] [PubMed]
63. Gryca, I.; Czerwińska, K.; Machura, B.; Chrobok, A.; Shul'pina, L.S.; Kuznetsov, M.L.; Nesterov, D.S.; Kozlov, Y.N.; Pombeiro, A.J.L.; Varyan, I.A.; et al. High catalytic activity of vanadium complexes in alkane oxidations with hydrogen peroxide: An effect of 8-hydroxyquinoline derivatives as noninnocent ligands. *Inorg. Chem.* **2018**, *57*, 1824–1839. [CrossRef] [PubMed]
64. Shiro, M.; Fernando, Q. Structures of two five-coordinated metal chelates of 2-methyl-8-quinolinol. *Anal. Chem.* **1971**, *43*, 1222–1230. [CrossRef]
65. Addison, A.W.; Rao, T.N. Synthesis, structure, and spectroscopic properties of copper(II) compounds containing nitrogen–sulphur donor ligands—The crystal and molecular structure of aqua [1,7-bis(N-methylbenzimidazol-2′-yl)-2,6-dithiaheptane]copper(II) perchlorate. *J. Chem. Soc. Dalton Trans.* **1984**, *1984*, 1349–1356. [CrossRef]
66. Groom, C.R.; Bruno, I.J.; Lightfoot, M.P.; Ward, S.C. The Cambridge structural database. *Acta Crystallogr. Sect. B Struct. Sci. Cryst. Eng. Mater.* **2016**, *72*, 171–179. [CrossRef]

67. Menati, S.; Rudbari, H.A.; Khorshidifard, M.; Jalilian, F. A new oxovanadium(IV) complex containing an O,N-bidentate Schiff base ligand: Synthesis at ambient temperature, characterization, crystal structure and catalytic performance in selective oxidation of sulfides to sulfones using H_2O_2 under solvent-free conditions. *J. Mol. Struct.* **2016**, *1103*, 94–102.
68. Grivani, G.; Khalaji, A.D.; Tahmasebi, V.; Gotoh, K.; Ishida, H. Synthesis, characterization and crystal structures of new bidentate Schiff base ligand and its vanadium(IV) complex: The catalytic activity of vanadyl complex in epoxidation of alkenes. *Polyhedron* **2012**, *31*, 265–271. [CrossRef]
69. Burgess, J.; Fawcett, J.; Palma, V.; Gilani, S.R. Fluoro derivatives of bis (salicylideneaminato-N, O) copper(II) and-oxovanadium(IV). *Acta Crystallogr. Sect. C Struct. Sci. Cryst. Eng. Mater.* **2001**, *57*, 277–280. [CrossRef]
70. Santoni, G.; Rehder, D. Structural models for the reduced form of vanadate-dependent peroxidases: Vanadyl complexes with bidentate chiral Schiff base ligands. *J. Inorg. Biochem.* **2004**, *98*, 758–764. [CrossRef]
71. Cornman, C.R.; Geiser-Bush, K.M.; Rowley, S.P.; Boyle, P.D. Structural and electron paramagnetic resonance studies of the square pyramidal to trigonal bipyramidal distortion of vanadyl complexes containing sterically crowded schiff base ligands. *Inorg. Chem.* **1997**, *36*, 6401–6408. [CrossRef]
72. Pasquali, M.; Marchetti, F.; Floriani, C.; Merlino, S. Oxovanadium(IV) complexes containing bidentate Schiff-base ligands: Synthesis and structural and spectroscopic data. *J. Chem. Soc. Dalton Trans.* **1977**, *1977*, 139–144. [CrossRef]
73. Cashin, B.; Cunningham, D.; Daly, P.; McArdle, P.; Munroe, M.; Chonchubhair, N.N. Donor properties of the vanadyl ion: Reactions of vanadyl salicylaldimine β-ketimine and acetylacetonato complexes with groups 14 and 15 Lewis acids. *Inorg. Chem.* **2002**, *41*, 773–782. [CrossRef] [PubMed]
74. Hsuan, R.E.; Hughes, J.E.; Miller, T.H.; Shaikh, N.; Cunningham, P.H.M.; O'Connor, A.E.; Tidey, J.P.; Blake, A.J. Crystal structure of {2,2'-[ethylenebis(nitrilomethanylylidene)]diphenolato-/4O,N,N',O'}oxidovanadium(IV) methanol monosolvate. *Acta Crystallogr. Sect. E Struct. Sci. Cryst. Eng. Mater* **2014**, *70*, m380–m381.
75. Nguyen, M.T.; Jones, R.A.; Holliday, B.J. Effect of conjugation length and metal-backbone interactions on charge transport properties of conducting metallopolymers. *Polym. Chem.* **2017**, *8*, 4359–4367. [CrossRef]
76. Carter, E.; Fallis, I.A.; Kariuki, B.M.; Morgan, I.R.; Murphy, D.M.; Tatchell, T.; Van Doorslaer, S.; Vinck, E. Structure and pulsed EPR characterization of N,N'-bis(5-tert-butylsalicylidene)-1,2-cyclohexanediamino-vanadium(IV) oxide and its adducts with propylene oxide. *Dalton Trans.* **2011**, *40*, 7454–7462. [CrossRef]
77. Hoshina, G.; Tsuchimoto, M.; Ohba, S.; Nakajima, K.; Uekusa, H.; Ohashi, Y.; Ishida, H.; Kojima, M. Thermal Dehydrogenation of Oxovanadium(IV) Complexes with Schiff Base Ligands Derived from meso-1,2-Diphenyl-1,2-ethanediamine in the Solid State. *Inorg. Chem.* **1998**, *37*, 142–145. [CrossRef]
78. Hoshina, G.; Tsuchimoto, M.; Ohba, S. exo-[(RS,SR)-N,N'-Bis(salicylidene)-2,3-butanediaminato]oxovanadium(IV). *Acta Crystallogr. Sect. C Struct. Sci. Cryst. Eng. Mater.* **1999**, *55*, 1082–1084. [CrossRef]
79. Bonadies, J.A.; Butler, W.M.; Pecoraro, V.L.; Carrano, C.J. Novel reactivity patterns of (N,N'-ethylenebis(salicylideneaminato))oxo vanadium(IV) in strongly acidic media. *Inorg. Chem.* **1987**, *26*, 1218–1222. [CrossRef]
80. Hoshina, G.; Tsuchimoto, M.; Ohba, S. endo-{6,6'-Diethoxy-2,2'-[(R)-propane-1,2-diylbis(nitrilomethylidene)]diphenolato-O,N,N',O'}oxovanadium(IV). *Acta Crystallogr. Sect. C Struct. Sci. Cryst. Eng. Mater.* **1999**, *55*, 1812–1813. [CrossRef]
81. Bolm, C.; Bienewald, F.; Harms, K. Syntheses and vanadium complex of salen-like bissulfoximines. *Synlett* **1996**, *8*, 775–776. [CrossRef]
82. Oyaizu, K.; Dewi, E.L.; Tsuchida, E. Coordination of BF_4^- to Oxovanadium(V) Complexes, Evidenced by the Redox Potential of Oxovanadium(IV/V) Couples in CH_2Cl_2. *Inorg. Chem.* **2003**, *42*, 1070–1075. [CrossRef] [PubMed]
83. Chasteen, N.D. *Biological Magnetic Resonance*; Berliner, L.J., Reuben, J., Eds.; Plenum: New York, NY, USA, 1981; Volume 3, pp. 53–119.
84. Velayutham, M.; Varghese, B.; Subramanian, S. Magneto–structural correlation studies of a ferromagnetically coupled dinuclear vanadium(IV) complex. Single-crystal EPR study. *Inorg. Chem.* **1998**, *37*, 1336–1340. [CrossRef] [PubMed]
85. Shulpin, B.; Attanasio, D.; Suber, L. Efficient H_2O_2 oxidation of alkanes and arenes to alkyl peroxides and phenols catalyzed by the system vanadate-pyrazine-2-carboxylic acid. *J. Catal.* **1993**, *142*, 147–152. [CrossRef]
86. Kirillov, A.M.; Shul'pin, G.B. Pyrazinecarboxylic acid and analogs: Highly efficient co-catalysts in the metal-complex-catalyzed oxidation of organic compounds. *Coord. Chem. Rev.* **2013**, *257*, 732–754. [CrossRef]
87. Levitsky, M.M.; Bilyachenko, A.N.; Shul'pin, G.B. Oxidation of CH compounds with peroxides catalyzed by polynuclear transition metal complexes in Si-or Ge-sesquioxane frameworks: A review. *J. Organomet. Chem.* **2017**, *849–850*, 201–218. [CrossRef]
88. Shul'pin, B.; Shul'pina, L.S. *Vanadium Catalysis*; Sutradhar, M., da Silva, J.A.L., Pombeiro, A.J.L., Eds.; Royal Society of Chemistry: London, UK, 2020; Chapter 4; pp. 72–96.
89. Shul'pin, G.B.; Süss-Fink, G. Oxidations by the reagent "H_2O_2–vanadium complex–pyrazine-2-carboxylic acid". Part 4. Oxidation of alkanes, benzene and alcohols by an adduct of H_2O_2 with urea. *J. Chem. Soc. Perkin Trans. 2* **1995**, *7*, 1459–1463. [CrossRef]
90. Shul'pin, G.B.; Drago, R.S.; Gonzalez, M. Oxidations by a "H_2O_2–vanadium complex–pyrazine-2-carboxylic acid" reagent. Part 5. Oxidation of lower alkanes with the formation of carbonyl compounds". *Russ. Chem. Bull.* **1996**, *45*, 2386–2388. [CrossRef]
91. Shul'pin, G.B.; Guerreiro, M.C.; Schuchardt, U. Oxidations by the reagent O_2–H_2O_2–vanadium complex–pyrazine-2-carboxylic acid. Part 7. Hydroperoxidation of higher alkanes. *Tetrahedron* **1996**, *52*, 13051–13062. [CrossRef]
92. Shul'pin, G.B.; Druzhinina, A.N.; Nizova, G.V. Oxidation with the H_2O_2–VO_3^-–pyrazine-2-carboxylic acid reagent. Part 2. Oxidation of alcohols and aromatic hydrocarbons. *Russ. Chem. Bull.* **1993**, *42*, 1327–1329.

93. Shul'pin, G.B.; Nesterov, D.S.; Shul'pina, L.S.; Pombeiro, A.J.L. A hydroperoxo-rebound mechanism of alkane oxidation with hydrogen peroxide catalyzed by binuclear manganese(IV) complex in the presence of an acid with involvement of atmospheric dioxygen. *Inorg. Chim. Acta* **2017**, *455*, 666–676. [CrossRef]
94. Ertik, O.; Kalındemirtaş, F.D.; Kaya, B.; Yanardag, R.; Erdem Kuruca, S.; Şahin, O.; Ülküseven, B. Oxovanadium(IV) complexes with tetradentate thiosemicarbazones. Synthesis, characterization, anticancer enzyme inhibition and in vitro cytotoxicity on breast cancer cells. *Polyhedron* **2021**, *202*, 115192. [CrossRef]
95. Sutradhar, M.; Alegria, E.; Ferretti, F.; Raposo, L.R.; Guedes da Silva, M.F.C.; Baptista, P.V.; Fernandes, A.R.; Pombeiro, A.J.L. Antiproliferative activity of heterometallic sodium and potassium-dioxidovanadium (V) polymers. *J. Inorg. Biochem.* **2019**, *200*, 110811. [CrossRef]
96. Yang, X.G.; Wang, K.; Lu, J.F.; Crans, D.C. Membrane transport of vanadium compounds and the interaction with the erythrocyte membrane. *Coord. Chem. Rev.* **2003**, *237*, 103–111. [CrossRef]
97. Reigosa-Chamorro, F.; Raposo, L.R.; Munin-Cruz, P.; Pereira, M.T.; Roma-Rodrigues, C.; Baptista, P.V.; Fernandes, A.R.; Vila, J.M. In Vitro and In Vivo Effect of Palladacycles: Targeting A2780 Ovarian Carcinoma Cells and Modulation of Angiogenesis. *Inorg. Chem.* **2021**, *60*, 3939–3951. [CrossRef] [PubMed]
98. Nycz, J.E.; Szala, M.; Malecki, G.J.; Nowak, M.; Kusz, J. Synthesis, spectroscopy and computational studies of selected hydroxyquinolines and their analogues. *Spectrochim. Acta Part A* **2014**, *117*, 351. [CrossRef]
99. *CrysAlisPRO*; Oxford Diffraction/Agilent Technologies UK Ltd.: Yarnton, UK, 2014.
100. Sheldrick, G.M. Crystal structure refinement with SHELXL. *Acta Crystallogr. Sect. C Struct. Sci. Cryst. Eng. Mater.* **2015**, *71*, 3–8. [CrossRef] [PubMed]
101. Sanna, D.; Várnagy, K.; Lihi, N.; Micera, G.; Garribba, E. Formation of new non-oxido vanadium(IV) species in aqueous solution and in the solid state by tridentate (O, N, O) ligands and rationalization of their EPR behavior. *Inorg. Chem.* **2013**, *52*, 8202–8213. [CrossRef]
102. *WINEPR SimFonia*, Version 1.25; Bruker Analytische, Messtechnik GmbH: Karlsruhe, Germany, 1996.
103. Raposo, L.R.; Silva, A.; Silva, D.; Roma-Rodrigues, C.; Espadinha, M.; Baptista, P.V.; Santos, M.M.M.; Fernandes, A.R. Exploiting the antiproliferative potential of spiropyrazoline oxindoles in a human ovarian cancer cell line. *Bioorg. Med. Chem.* **2021**, *30*, 115880. [CrossRef]
104. Choroba, K.; Machura, B.; Szlapa-Kula, A.; Malecki, J.G.; Raposo, L.; Roma-Rodrigues, C.; Cordeiro, S.; Baptista, P.V.; Fernandes, A.R. Square planar Au (III), Pt (II) and Cu (II) complexes with quinoline-substituted 2, 2′: 6′, 2″-terpyridine ligands: From in vitro to in vivo biological properties. *Eur. J. Med. Chem.* **2021**, *218*, 113404. [CrossRef] [PubMed]
105. Reers, M.; Smith, T.W.; Chen, L.B. J-aggregate formation of a carbocyanine as a quantitative fluorescent indicator of membrane potential. *Biochemistry* **1991**, *30*, 4480–4486. [CrossRef] [PubMed]
106. Shul'pin, G.B. Metal-catalyzed hydrocarbon oxygenations in solutions: The dramatic role of additives: A review. *J. Mol. Catal. A Chem.* **2002**, *189*, 39–66. [CrossRef]
107. Shul'pin, G.B.; Kozlov, Y.N.; Shul'pina, L.S.; Petrovskiy, P.V. Oxidation of alkanes and alcohols with hydrogen peroxide catalyzed by complex Os3(CO)10(μ-H)2. *Appl. Organometal. Chem.* **2010**, *24*, 464–472. [CrossRef]

Article

A Zinc-Mediated Deprotective Annulation Approach to New Polycyclic Heterocycles

Lucia Veltri [1,*], Roberta Amuso [1], Marzia Petrilli [1], Corrado Cuocci [2], Maria A. Chiacchio [3], Paola Vitale [4] and Bartolo Gabriele [1,*]

[1] Laboratory of Industrial and Synthetic Organic Chemistry (LISOC), Department of Chemistry and Chemical Technologies, University of Calabria, Via Pietro Bucci 12/C, 87036 Arcavacata di Rende, Italy; robyamuso@gmail.com (R.A.); marzia_p94@hotmail.it (M.P.)
[2] Institute of Crystallography, National Research Council, Via Amendola, 122/O, 70126 Bari, Italy; corrado.cuocci@ic.cnr.it
[3] Department of Drug Sciences, University of Catania, Viale A. Doria 6, 95125 Catania, Italy; ma.chiacchio@unict.it
[4] Department of Pharmacy—Pharmaceutical Sciences, University of Bari "Aldo Moro", Via E. Orabona 4, 70125 Bari, Italy; paola.vitale@uniba.it
* Correspondence: lucia.veltri@unical.it (L.V.); bartolo.gabriele@unical.it (B.G.); Tel.: +39-0984-492817 (L.V.); +39-0984-492815 (B.G.)

Abstract: A straightforward approach to new polycyclic heterocycles, 1*H*-benzo[4,5]imidazo[1,2-*c*][1,3]oxazin-1-ones, is presented. It is based on the ZnCl$_2$-promoted deprotective 6-*endo-dig* heterocyclization of *N*-Boc-2-alkynylbenzimidazoles under mild conditions (CH$_2$Cl$_2$, 40 °C for 3 h). The zinc center plays a dual role, as it promotes Boc deprotection (with formation of the *tert*-butyl carbocation, which can be trapped by substrates bearing a nucleophilic group) and activates the triple bond toward intramolecular nucleophilic attack by the carbamate group. The structure of representative products has been confirmed by X-ray diffraction analysis.

Keywords: alkynes; annulation; benzimidazoxazinones; heterocycles; polycyclic heterocycles; heterocyclization; zinc

1. Introduction

The development of efficient methods for the synthesis of high value added polycyclic heterocyclic derivatives by metal-promoted annulation of acyclic precursors is one of the most important area of research in heterocyclic chemistry [1–5]. Polycyclic heterocyclic systems, in fact, are largely present as fundamental cores in natural products and in biologically active compounds [6–11], and the possibility to obtain them by a simple cyclization process starting from readily available substrates is particularly attractive [1–5].

Among acyclic substrates able to undergo a metal-promoted cyclization to give a polycyclic heterocycle, functionalized alkynes bearing a suitably placed heteronucleophile play a major role, as the triple bond can be easily electrophilically activated by a suitable metal species thus promoting the cyclization by intramolecular nucleophilic attack [1–5]. Usually, processes like these are promoted by costly metals (mainly gold [12–19], palladium [20–23], rhodium [24–26], platinum [27–29], and, occasionally, ruthenium [30]), while the use of less expensive metal species, such as cobalt [31], nickel [32], copper [33–36], zinc [37–40], and silver [41,42] compounds, has been scantly reported in the literature, and applied to a limited number of examples.

In this work, we report on the use of very simple and inexpensive ZnCl$_2$ as a promoter for the efficient deprotective heterocyclization of *N*-Boc-2-alkynylbenzimidazoles **1**, to give access to novel polycyclic heterocycles, that are, 1*H*-benzo[4,5]imidazo[1,2-*c*][1,3]oxazin-1-ones **2** (Scheme 1). It is worth mentioning in this context that the cyclization of *O*-Boc propargyl alcohols to give 4*H*-1,3-dioxin-2-ones and/or 4-alkylidene-1,3-dioxolan-2-ones has been previously reported to occur with mercuric triflate as the catalyst [43]. It is also important to note that some excellent reviews on Zn-catalyzed reactions have appeared in the recent literature [44–48].

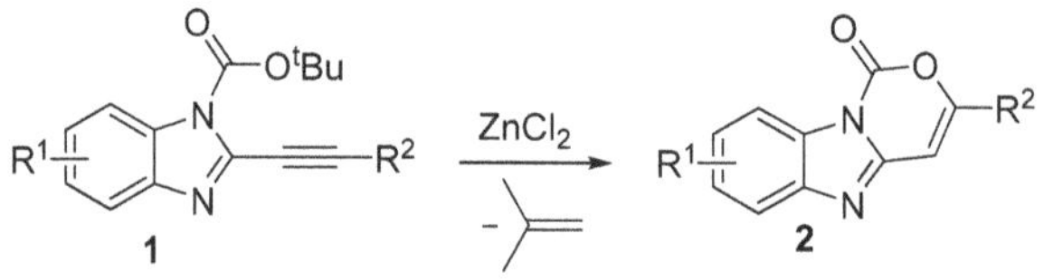

Scheme 1. This work: $ZnCl_2$-assisted heterocyclization of N-Boc-alkynylbenzimidazoles **1** to benzimidazoxaxinones **2**.

2. Results and Discussion

It is well known that zinc (II) compounds are able to promote Boc deprotection [49–54]. In particular, an excess of $ZnBr_2$ has been successfully employed for the deprotection of N-Boc secondary amines [52] as well as of *tert*-butyl esters [53,54]. Considering the importance of developing new approaches to the synthesis of polycyclic heterocycles by heterocyclization processes promoted by non-noble and inexpensive metal species, we have explored the possibility to access new polycyclic heterocycles, that are 1H-benzo[4,5]imidazo[1,2-c][1,3]oxazin-1-ones **2**, starting from readily available N-Boc-2-alkynylbenzimidazoles **1**, by Zn(II)-assisted deprotective heterocyclization (Scheme 1). According to our rationale, the zinc center should play a double role, that is, to promote deprotection to give a carbamate species **A** (with elimination of isobutene and H^+ from the ensuing *tert*-butyl carbocation [52–54]) and then assist a 6-*endo-dig* heterocyclization by intramolecular nucleophilic attack of the free carbamate group of species **B** (in equilibrium with **A**) on the triple bond activated by coordination to Zn^{2+} (with the zinc center stabilized by chelation by the benzimidazole nitrogen). This would lead to organizinc intermediate **C**, whose protonolysis would then afford the polycyclic heterocycles **2** (Scheme 2; zinc counteranions have been omitted for clarity).

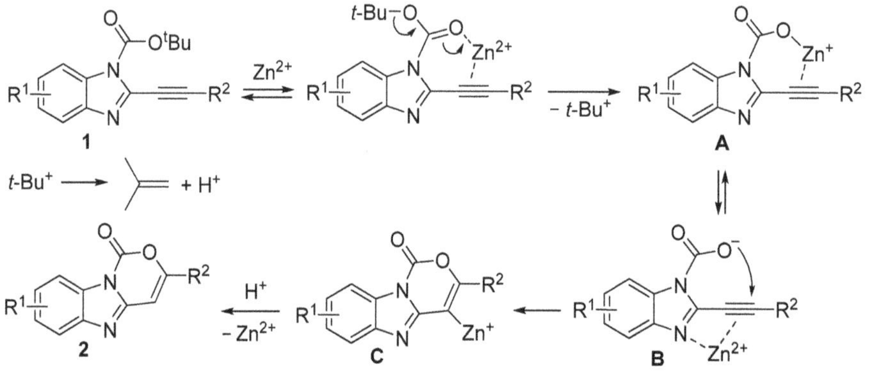

Scheme 2. Mechanistic hypothesis for the formation of polycyclic heterocycles **2** by Zn^{2+}-mediated sequential deprotection - 6-*endo-dig* heterocyclization of N-Boc-alkynylbenzimidazoles **1**.

The first experiments were performed using N-Boc-2-(hex-1-in-1-yl)-1H-benzo[d]imidazole **1a** as substrate (R^1 = H, R^2 = Bu) (prepared by alkynylation of N-Boc-2-bromo-1H-benzo[d]imidazole, see the Supplementary Materials for details), which was allowed to react in CH_2Cl_2 as the solvent at room temperature in the presence of $ZnBr_2$ (1 equiv). Under these conditions, after 3 h reaction time, substrate conversion was 51%, while the desired 3-butyl-1H-benzo[4,5]imidazo[1,2-c][1,3]oxazin-1-one **2a** was isolated in 25% yield. The structure of **2a** was unequivocally confirmed by XRD analysis (see the

Supplementary Materials for XRD data). The X-ray structure of **2a**, shown in Figure 1, confirmed that the heterocyclization process at intermediate **B** level occurred in a 6-*endo-dig* fashion (with closure to a 6-membered ring) rather than in the possible alternative 5-*exo-dig* fashion (with closure to a five-membered ring).

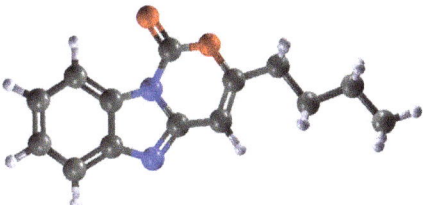

Figure 1. Molecular structure of 3-butyl-1*H*-benzo[4,5]imidazo[1,2-*c*][1,3]oxazin-1-one **2a**. Color legend: carbon (light grey), hydrogen (white), oxygen (red), nitrogen (blue) (CCDC 2050576).

In spite of the low yield, this initial result was encouraging, since it confirmed the validity of our work hypothesis and the possibility to synthesize novel polycyclic heterocycles with a very simple approach and using an inexpensive promoter. In order to improve the reaction performance, and achieve a higher **2a** yield, we then changed some operative parameters (Table 1, entries 2–9). Practically no reaction occurred by changing the solvent to MeOH (Table 1, entry 2), while only traces of **2a** were detected in acetone (Table 1, entry 3). Lowering the amount of $ZnBr_2$ significantly suppressed the reaction (Table 1, entry 4). On the other hand, the use of 1.5 or 2 equiv of $ZnBr_2$ was beneficial, **2a** being formed in ca. 70% isolated yield (Table 1, entries 5 and 6, respectively). Better results with respect to the parent reaction (Table 1, entry 1) were also obtained by increasing the **1a** concentration from 0.5 (Table 1, entry 1) to 1 mmol/mL of CH_2Cl_2 (Table 1, entry 7), while more diluted conditions led to a lower **2a** yield (Table 1, entry 8). Predictably, a faster reaction was observed at 40 °C rather than 25 °C, with a higher yield of **2a** (Table 1, entry 9) with respect to the initial experiment (Table 1, entry 1). Under the optimized conditions (40 °C in CH_2Cl_2 in the presence of 1.5 equiv of $ZnBr_2$, with a substrate concentration of 1 mmol per mL of solvent), **2a** could be finally obtained in a yield as high as 79% (Table 1, entry 10).

Very interestingly, the reaction was also successful using $ZnCl_2$ (Table 1, entry 11) or ZnI_2 (Table 1, entry 12), the best results in terms of **2a** yield being obtained with $ZnCl_2$ (82%, Table 1, entry 11). This result, associated with the lower cost of $ZnCl_2$, made $ZnCl_2$ the promoter of choice for realizing the transformation of **1a** into benzimidazoxazinone **2a** and for the subsequent extension to other differently substituted substrates (Table 2). Thus, to assess the generality of the reaction, various *N*-Boc-alkynylbenzimidazoles **1** (bearing different R^1 and R^2 groups; prepared as detailed in the Supplementary Materials) were subjected to the optimized reaction conditions with $ZnCl_2$ as the promoter (Table 2, entries 2–15).

Table 1. ZnX$_2$-promoted deprotective heterocyclization of *N*-Boc-2-(hex-1-in-1-yl)-1*H*-benzo[*d*]imidazole **1a** under different conditions [a].

Entry	ZnX$_2$ (Equiv)	T (°C)	Solvent	Concentration of 1a [b]	Conversion of 1a (%) [c]	Yield of 2a (%) [d]
1	ZnBr$_2$ (1)	25	CH$_2$Cl$_2$	0.5	51	25
2	ZnBr$_2$ (1)	25	MeOH	0.5	3	0
3	ZnBr$_2$ (1)	25	acetone	0.5	12	Traces
4	ZnBr$_2$ (0.5)	25	CH$_2$Cl$_2$	0.5	9	6
5	ZnBr$_2$ (1.5)	25	CH$_2$Cl$_2$	0.5	100	72
6	ZnBr$_2$ (2)	25	CH$_2$Cl$_2$	0.5	100	70
7	ZnBr$_2$ (1)	25	CH$_2$Cl$_2$	1.0	62	33
8	ZnBr$_2$ (1)	25	CH$_2$Cl$_2$	0.2	42	10
9	ZnBr$_2$ (1)	40	CH$_2$Cl$_2$	0.5	100	63
10	ZnBr$_2$ (1.5)	40	CH$_2$Cl$_2$	1.0	100	79
11	ZnCl$_2$ (1.5)	40	CH$_2$Cl$_2$	1.0	100	82
12	ZnI$_2$ (1.5)	40	CH$_2$Cl$_2$	1.0	100	77

[a] All reactions were carried out for 3 h. [b] Mmol of starting **1a** per mL of solvent. [c] Based on unreacted **1a**, upon isolation from the reaction mixture. [d] Isolated yield based on starting **1a**.

Table 2. Synthesis of 1*H*-benzo[4,5]imidazo[1,2-*c*][1,3]oxazin-1-ones **2** by ZnCl$_2$-promoted deprotective heterocyclization of *N*-Boc-2-alkynylbenzimidazoles **1** [a].

Entry	1	2	Yield of 2 (%) [b]
1	1a	2a	82
2	1b	2b	77
3	1c	2c	76
4	1d	2d	83

Table 2. Cont.

Entry	1	2	Yield of 2 (%) [b]
5	1e	2e	77
6	1f	2f	45 [c]
7	1g	2g	30 [d]
8	1h	2h	85
9	1i	2i	82
10	1j	2j	80
11	1k	2k	70
12	1l	2l	66
13	1m	2m	60
14	1n	2n	74

Table 2. Cont.

Entry	1	2	Yield of 2 (%) [b]
15	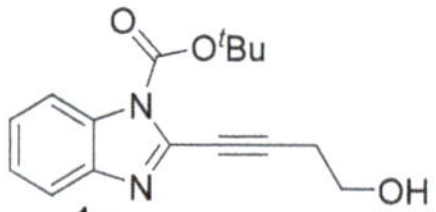 **1o**: benzimidazole with N-C(O)-O^{t}Bu, 2-substituted with C≡C-CH$_2$-CH$_2$-OH	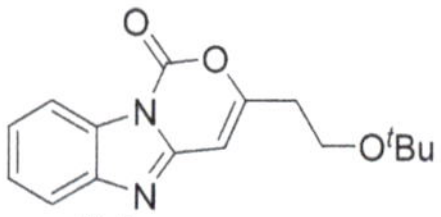 **2o'**: benzo-fused oxazinone with CH$_2$CH$_2$-O^{t}Bu substituent	66

[a] All reactions were carried out in CH$_2$Cl$_2$ (1 mmol of **1** per mL of solvent) at 40 °C for 3 h. [b] Isolated yield based on starting **1**. [c] The reaction led also to 2-(hex-1-yn-1-yl)-6-nitro-1H-benzo[d]imidazole **3f** in 20% isolated yield. [d] The reaction led also to 2-(hex-1-yn-1-yl)-5-nitro-1H-benzo[d]imidazole **3g** in 31% isolated yield.

As can be seen from Table 2, entries 2–5, excellent results were obtained with substrates still with R^2 = Bu and bearing either electron-donating (methyl or methoxy; yields of the corresponding products **2b–d** were 76–83%, Table 2, entries 2–4) or electron-withdrawing chlorine substituents (yield of **2e** = 77%, Table 2, entry 5) on the aromatic ring. On the other hand, inferior results were observed with substrates **1f** and **1g**, bearing a strong electron-withdrawing nitro substituent (yields of **2f** and **2g** were 45% and 30%, Table 2, entries 6 and 7, respectively). With these substrates, complete Boc removal competed with heterocyclization, as confirmed by the formation of not negligible amounts of deprotected compounds **3f** and **3g** (20% and 31%, respectively, Table 2, entries 6 and 7) (Scheme 3), not observed in other cases. Clearly, the formation of these byproducts from substrates **1f** and **1g** is due to the diminished nucleophilicity of the carbamate intermediate **B** (Scheme 2) caused by the strong electron-withdrawing effect of the nitro group, which makes decarboxylation to compete with cyclization. The structures of products **2c** and **2f** were confirmed by XRD analysis (see the Supplementary Materials for XRD data). The X-ray structures of **2c** and **2f**, shown in Figures 2 and 3, respectively, allowed to unequivocally establish the positions of the methoxy and nitro substituents in regioisomeric substrates **1c/1d** and **1f/1g**, respectively (as **2c** must be formed from **1c** and **2f** from **1f**).

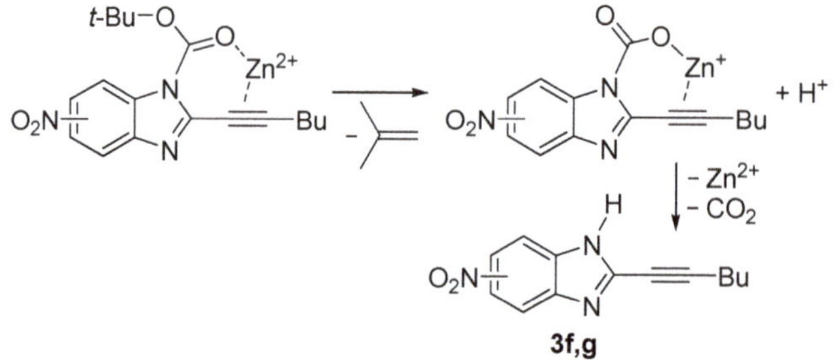

Scheme 3. Formation of byproducts **3f** and **3g** (Table 2, entries 6 and 7) by Boc deprotection of nitro-substituted substrates **1f** and **1g**, competitive with heterocyclization.

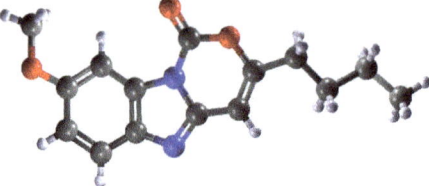

Figure 2. Molecular structure of 3-butyl-8-methoxy-1H-benzo[4,5]imidazo[1,2-c][1,3]oxazin-1-one **2c**. Color legend: carbon (light grey), hydrogen (white), oxygen (red), nitrogen (blue) (CCDC 2051334).

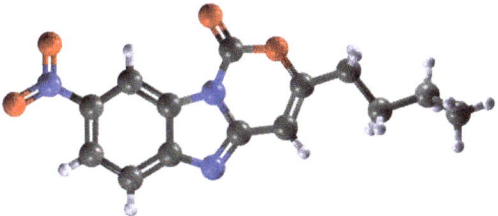

Figure 3. Molecular structure of 3-butyl-8-nitro-1*H*-benzo[4,5]imidazo[1,2-*c*][1,3]oxazin-1-one **2f**. Color legend: carbon (light grey), hydrogen (white), oxygen (red), nitrogen (blue) (CCDC 2050711).

High yields of the corresponding benzimidazoxazinones were obtained by changing the alkyl substituent on the triple bond R^2 to octyl (yield of **2h**, 85%; Table 2, entry 8), isopentyl (yield of **2i**, 82%; Table 2, entry 9), or phenethyl (yield of **2j**, 80%; Table 2, entry 10), while a slightly lower yield was observed with R^2 = cyclohexylmethyl (yield of **2k**, 70%; Table 2, entry 11). The use of a substrate with the triple bond conjugated with an alkenyl group, as in *N*-Boc-2-(cyclohex-1-en-1-ylethynyl)-1*H*-benzo[*d*]imidazole **1l**, led to a satisfactory yield of the corresponding polycyclic heterocycle **2l** (66%; Table 2, entry 12).

The method also worked nicely with substrates bearing a functionalized alkyl chain of the triple bond, as shown by the results obtained with a methoxymethyl (yield of **2m**, 60%; Table 2, entry 13) or a 2-(methoxycarbonyl) ethyl (yield of **2n**, 74%; Table 2, entry 14) group. Interestingly in the case of *N*-Boc-4-(1*H*-benzo[*d*]imidazol-2-yl) but-3-yn-1-ol **1o**, bearing a 2-hydroxyethyl group on the triple bond, the *tert*-butyl group was incorporated into the final product to give 3-(2-(*tert*-butoxy)ethyl)-1*H*-benzo[4,5]imidazo[1,2-*c*][1,3]oxazin-1-one **2o'** (66% yield; Table 1, entry 15). This is clearly due to the trapping of the *tert*-butyl carbocation, ensuing from deprotection, by the nucleophilic hydroxyl group, as shown in Scheme 4.

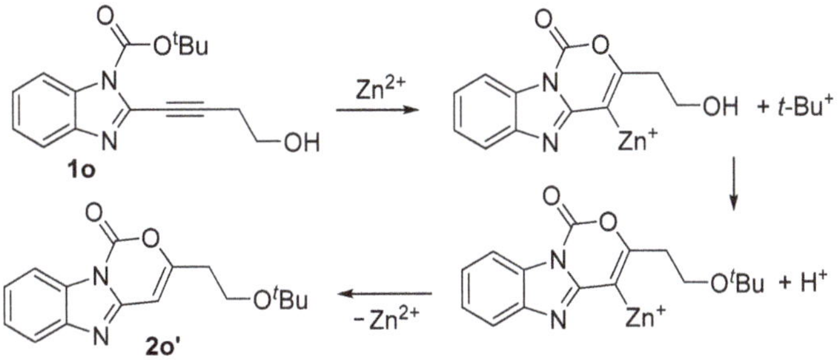

Scheme 4. Plausible mechanism for the formation of product **2o'** (chloride anions are omitted for clarity).

3. Materials and Methods

3.1. General Experimental Methods

Melting points were measured with a Leitz Laborlux 12 POL polarizing optical microscope (Leitz Italia GmbH/Srl, Lana(BZ), Italy) and are uncorrected. ^{1}H NMR and ^{13}C NMR spectra were recorded at 25 °C in CDCl$_3$ or DMSO-d_6 at 300 MHz or 500 MHz and 75 or 125 MHz, respectively, with Me$_4$Si as internal standard, using Bruker DPX Avance 300 and Bruker DPX Avance 500 NMR spectrometers (Brucker Italia s.r.l., Milano, Italy); chemical shifts (δ) and coupling constants (*J*) are given in ppm and in Hz, respectively. IR spectra were taken with a JASCO FT-IR 4200 spectrometer (Jasco Europe s.r.l., Cremella, Lecco, Italy). All reactions were analyzed by TLC on silica gel 60 F$_{254}$ and by GC-MS using a Shimadzu QP-2010 GC–MS apparatus (Smimadzu Italia s.r.l., Milano, Italy) at 70 eV ionization voltage equipped with a 95% methyl polysiloxane–5% phenyl polysiloxane capillary column (30 m × 0.25 mm, 0.25 µm). Column chromatography was performed on silica gel 60 (Merck, 70–230 mesh; Merck Life Science s.r.l., Milano, Italy). Evaporation refers to the removal of solvent under reduced pressure. The HRMS spectra were taken on an Agilent 1260 Infinity UHD accurate-mass Q-TOF mass spectrometer (Agilent Technologies Italia s.p.a. Cernusco sul Naviglio, Milano, Italy), equipped with an electrospray ion source (ESI) operated in dual ion mode. Ten microliters of the sample solutions (CH$_3$OH) were introduced by continuous infusion at a flow rate of 200 L min^{-1} with the aid of a syringe pump. Experimental conditions were performed as follows: capillary voltage, 4000 V; nebulizer pressure, 20 psi; flow rate of drying gas, 10 L/min; temperature of sheath gas, 325 °C; flow rate of sheath gas, 10 L/min; skimmer voltage, 60 V; OCT1 RF Vpp, 750 V; fragmentor voltage, 170 V. The spectra data were recorded in the m/z range of 100–1000 Da in a centroid pattern of full-scan MS analysis mode. The MS/MS data of the selected compounds were obtained by regulating diverse collision energy (18–45 eV).

3.2. Preparation of Substrates **1**

Substrates were prepared and characterized as described in the Supplementary Materials.

3.3. General Procedure for the Synthesis of Benzimidazoxazinone Derivatives **2**

See Table 2 for reference. A Schlenk flask was charged under nitrogen with the *N*-Boc-2-alkynylbenzimidazole **1** (1 mmol) (**1a**: 298 mg; **1b**: 326 mg; **1c**: 328 mg; **1d**: 328 mg; **1e**: 367 mg; **1f**: 343 mg; **1g**: 343 mg; **1h**: 354 mg; **1i**: 312 mg; **1j**: 346 mg; **1k**: 338 mg; **1l**: 322 mg, **1m**: 286 mg, **1n**: 328 mg; **1o**: 286 mg), anhydrous CH$_2$Cl$_2$ (1 mL), and ZnCl$_2$ (204 mg, 1.5 mmol). The reaction mixture was heated at 40 °C and then allowed to stir at this temperature for 3 h. After cooling, the reaction mixture was diluted with CH$_2$Cl$_2$ (5 mL) and water (5 mL) (for **2a-l**, **2n**, and **2o'**). Alternatively, after cooling, the solvent was evaporated, and water (20 mL) was added to the residue (for **2m**). Phases were separated the aqueous phase was washed with CH$_2$Cl$_2$ (5 mL), and the combined organic phases were dried with Na$_2$SO$_4$. After filtration and evaporation of the solvent, the product was purified by column chromatography on silica gel using hexane/AcOEt (8:2, *v/v*) as the eluent (for **2a-1l**, **2n**, and **2o'**). For the purification of **2m**, the suspension obtained as seen above was filtered, the precipitate washed with water (3 × 5 mL) and then purified by column chromatography on silica gel using hexane/AcOEt (8:2, *v/v*) as eluent. With substrates **1f** and **1g**, the reaction also led to the formation of deprotected products **3f** and **3g**, respectively (Scheme 3) (order of elution: **3f** followed by **2f**; **2g** followed by **3g**).

3.3.1. 3-Butyl-1*H*-benzo[4,5]imidazo[1,2-*c*][1,3]oxazin-1-one **2a**

Yield: 198 mg, starting from 298 mg of **1a** (82%) (Table 2, entry 1). Colorless solid, mp: 92–94 °C; IR (KBr): ν = 1759 (s), 1667 (m), 1551 (w), 1450 (w), 1366 (s), 1096 (m), 972 (w), 849 (w), 748 (m) cm^{-1}; ^{1}H NMR (300 MHz, CDCl$_3$) δ = 8.24–8.13 (m, 1 H, aromatic), 7.82–7.73 (m, 1 H, aromatic), 7.52–7.36 (m, 2 H, aromatic), 6.50 (s, 1 H, H-4), 2.61 (t, *J* = 7.3, 2 H, =CCH$_2$), 1.75 (quint, *J* = 7.3, 2 H, CH$_2$CH$_2$CH$_3$), 1.46 (hexuplet, *J* = 7.3, 2 H, CH$_2$CH$_3$), 0.98 (t, *J* = 7.3, 3 H, CH$_3$); ^{13}C NMR (75 MHz, CDCl$_3$): δ = 162.9, 147.4, 144.1, 129.3, 126.3, 124.9, 119.7, 114.6, 96.6, 32.8, 28.4, 22.1, 13.7; GC/MS = 242 (M$^+$, 100), 227 (2), 213 (3), 200 (42), 185 (31), 171 (6), 158 (43); 144 (4), 130 (12); HRMS (ESI-TOF) *m/z*: [M + H]$^+$ Calcd for C$_{14}$H$_{15}$N$_2$O$_2$$^+$ 243.1128; Found: 243.1132.

3.3.2. 3-Butyl-7,8-dimethyl-1*H*-benzo[4,5]imidazo[1,2-*c*][1,3]oxazin-1-one **2b**

Yield: 208 mg, starting from 326 mg of **1b** (77%) (Table 2, entry 2). Colorless solid, mp: 133–137 °C; IR (KBr): ν = 1768 (s), 1667 (m), 1558 (w), 1450 (m), 1381 (s), 1111 (w), 741 (w) cm^{-1}; ^{1}H NMR (300 MHz, CDCl$_3$): δ = 7.92 (s, 1 H, H-6 or H-9), 7.48 (s, 1 H, H-9 or H-6), 6.44 (s, 1 H, H-4), 2.59 (t, *J* = 7.5, 2 H, =CCH$_2$), 2.40 (s, 3 H, CH$_3$ at C-7 or C-8), 2.38 (s, 3 H, CH$_3$ at C-8 or C-7), 1.72 (quint, *J* = 7.5, 2 H, CH$_2$CH$_2$CH$_3$), 1.44 (hexuplet, *J* = 7.5, 2 H, CH$_2$CH$_3$), 0.98 (t, *J* = 7.5, 3 H, CH$_3$); ^{13}C NMR (75 MHz, CDCl$_3$): δ = 162.0, 146.6, 144.2, 142.5, 135.3, 134.3, 127.6, 119.8, 114.7, 96.7, 32.8, 28.5, 22.1, 20.4, 13.7; GC/MS = 270 (M$^+$,

100); 255 (3), 228 (29), 213 (24), 199 (5), 186 (19), 172 (3), 158 (6), 143 (1), 130 (2), 118 (8); HRMS (ESI-TOF) m/z: [M + H]$^+$ Calcd for $C_{16}H_{19}N_2O_2^+$ 271.1441; Found: 271.1446.

3.3.3. 3-Butyl-8-methoxy-1H-benzo[4,5]imidazo[1,2-c][1,3]oxazin-1-one **2c**

Yield: 207 mg, starting from 328 mg of **1c** (76%) (Table 2, entry 3). Colorless solid, mp: 96–99 °C; IR (KBr): ν = 1767 (s), 1667 (m), 1489 (m), 1443 (w), 1366 (m), 1281 (m), 1204 (w), 1026 (w), 818 (m) cm^{-1}; ^{1}H NMR (500 MHz, CDCl$_3$) δ = 7.70 (d, J = 2.5, 1 H, H-9), 7.63 (d, J = 8.8, 1 H, H-6), 7.06 (dd, J = 8.8, 2.5, 1 H, H-7), 6.46–6.44 (m, 1 H, H-4), 2.60 (t, J = 7.5, 2 H, =CCH$_2$), 1.72 (quint, J = 7.5, 2 H, CH$_2$CH$_2$CH$_3$), 1.45 (hexuplet, J = 7.5, 2 H, CH$_2$CH$_3$), 0.98 (t, J = 7.5, 3 H, CH$_3$); ^{13}C NMR (125 MHz, CDCl$_3$): δ = 161.7, 158.0, 146.3, 144.4, 138.3, 130.2, 120.2, 115.5, 98.3, 96.8, 56.0, 32.8, 28.5, 22.1, 13.7; GC/MS: m/z = 272 (M$^+$, 100), 257 (17), 229 (29), 215 (22), 187 (14); HRMS (ESI-TOF) m/z: [M + H]$^+$ Calcd for $C_{15}H_{17}N_2O_3^+$ 273.1234; Found: 273.1237.

3.3.4. 3-Butyl-7-methoxy-1H-benzo[4,5]imidazo[1,2-c][1,3]oxazin-1-one **2d**

Yield: 226 mg, starting from 328 mg of **1d** (83%) (Table 2, entry 4) Colorless solid, mp: 93–97 °C; IR (KBr): ν = 1760 (s), 1659 (m), 1558 (w), 1489 (m), 1435 (w), 1366 (m), 1281 (m), 1150 (m), 1103 (m) cm^{-1}; ^{1}H NMR (500 MHz, CDCl$_3$) δ = 8.06 (d, J = 8.9, 1 H, H-9), 7.24 (s, br, 1 H, H-6), 7.06–7.00 (m, 1 H, H-8), 6.50–6.47 (m, 1 H, H-4), 2.61 (t, J = 7.5, 2 H, =CCH$_2$), 1.73 (quint, J = 7.5, 2 H, CH$_2$CH$_2$CH$_3$), 1.45 (hexuplet, J = 7.5, 2 H, CH$_2$CH$_3$), 0.98 (t, J = 7.5, 3 H, CH$_3$); ^{13}C NMR (125 MHz, CDCl$_3$): δ = 162.7, 158.9, 148.0, 145.5, 144.0, 123.5, 114.9, 113.8, 102.7, 96.6, 55.8, 32.8, 28.5, 22.1, 13.7; GC/MS: m/z = 272 (M$^+$, 100), 230 (20), 215 (15), 199 (11), 188 (19); HRMS (ESI-TOF) m/z: [M + H]$^+$ Calcd for $C_{15}H_{17}N_2O_3^+$ 273.1234; Found: 273.1242.

3.3.5. 3-Butyl-7,8-dichloro-1H-benzo[4,5]imidazo[1,2-c][1,3]oxazin-1-one **2e**

Yield: 240 mg, starting from 367 mg of **1e** (77%) (Table 2, entry 5). Colorless solid, mp: 143–147 °C. IR (KBr): ν = 1775 (s), 1667 (m), 1543 (w), 1435 (w), 1350 (m), 1134 (w), 1096 (m) cm^{-1}; ^{1}H NMR (500 MHz, DMSO-d_6) δ = 8.18 (s, 1 H, H-6 or H-9), 8.07 (s, 1 H, H-9 or H-6), 6.91 (s, 1 H, H-4), 2.63 (t, J = 7.4, 2 H, =CCH$_2$), 1.65 (quint, J = 7.4, 2 H, =CCH$_2$CH$_2$), 1.40 (hexuplet, J = 7.4, 2 H, CH$_2$CH$_3$), 0.93 (t, J = 7.4, 3 H, CH$_3$); ^{13}C NMR (125 MHz, DMSO-d_6): δ = 163.9, 150.0, 143.5, 128.7, 128.3, 126.3, 120.5, 114.9, 96.3, 31.8, 27.9, 21.3, 13.5; GC/MS: m/z = 312 [(M + 2)$^+$, 61], 310 (M$^+$, 100), 268 (25), 253 (22), 226 (31), 202 (6); HRMS (ESI-TOF) m/z: [M + H]$^+$ Calcd for $C_{14}H_{13}Cl_2N_2O_2^+$ 311.0349; Found: 311.0348.

3.3.6. 3-Butyl-8-nitro-1H-benzo[4,5]imidazo[1,2-c][1,3]oxazin-1-one **2f**

Yield: 129 mg, starting from 343 mg of **1f** (45%) (Table 2, entry 6). Colorless solid, mp: 165–168 °C; IR (KBr): ν = 1775 (s), 1659 (m), 1543 (m), 1520 (m), 1343 (m), 748 (w) cm^{-1}; ^{1}H NMR (300 MHz, CDCl$_3$) δ = 9.05 (d, J = 1.9, 1 H, H-9), 8.39 (dd, J = 8.8, 1.9, 1 H, H-7), 7.83 (d, J = 8.8, 1 H, H-6), 6.63 (s, 1 H, H-4), 2.70 (t, J = 7.4, 2 H, =CCH$_2$), 1.76 (quint, J = 7.4, 2 H, CH$_2$CH$_2$CH$_3$), 1.49 (hexuplet, J = 7.4, 2 H, CH$_2$CH$_3$), 1.00 (t, J = 7.4, 3 H, CH$_3$); ^{13}C NMR (75 MHz, CDCl$_3$): δ = 165.6, 151.5, 148.6, 144.7, 143.2, 129.0, 122.1, 119.8, 111.0, 96.6, 33.1, 28.4, 22.1, 13.7; GC/MS: m/z = 287 (M$^+$, 100), 257 (11), 245 (49), 230 (27), 203 (23), 184 (16); HRMS (ESI-TOF) m/z: [M + Na + MeOH]$^+$ Calcd for $C_{15}H_{17}N_3O_5Na^+$ 342.1060; Found: 342.1064.

3.3.7. 3-Butyl-7-nitro-1H-benzo[4,5]imidazo[1,2-c][1,3]oxazin-1-one **2g**

Yield: 86 mg, starting from 343 mg of **1g** (30%) (Table 2, entry 7). Yellow solid, mp: 144–147 °C; IR (KBr): 1775 (s), 1667 (m), 1520 (s), 1350 (s), 1173 (w), 1119 (w), 934 (w), 833 (m), 741 (m) cm^{-1}; ^{1}H NMR (300 MHz, CDCl$_3$) δ = 8.62 (s, 1 H, H-6), 8.38–8.30 (m, 2 H, H-8 + H-9), 6.60 (s, 1 H, H-4), 2.68 (t, J = 7.4, 2 H, =CCH$_2$), 1.76 (quint, J = 7.4, 2 H, CH$_2$CH$_2$CH$_3$), 1.49 (hexuplet, J = 7.4, 2 H, CH$_2$CH$_3$), 1.00 (t, J = 7.4, 3 H, CH$_3$); ^{13}C NMR (75 MHz, DMSO-d_6): δ = 164.7, 150.1, 146.4, 144.2, 143.5, 133.5, 120.2, 115.8, 114.7, 96.5, 33.0,

28.4, 22.1, 13.7; GC/MS: m/z = 287 (M⁺, 100), 245 (50), 230 (29), 203 (31), 184 (13); HRMS (ESI-TOF) m/z: [M + Na + MeOH]⁺ Calcd for $C_{15}H_{17}N_3O_5Na^+$ 342.1060; Found: 342.1064.

3.3.8. 3-Octyl-1H-benzo[4,5]imidazo[1,2-c][1,3]oxazin-1-one **2h**

Yield: 254 mg, starting from 354 mg of **1h** (85%) (Table 2, entry 8). Colorless solid, mp: 90–94 °C; IR (KBr): ν = 1759 (s), 1667 (m), 1551 (m), 1396 (m), 1373 (m), 1134 (m), 1103 (m), 964 (w), 756 (m) cm⁻¹; ¹H NMR (300 MHz, CDCl₃): δ = 8.24–8.17 (m, 1 H, aromatic), 7.83–7.75 (m, 1 H, aromatic), 7.50–7.39 (m, 2 H, aromatic), 6.70 (s, 1 H, H-4), 2.62 (t, J = 7.6, 2 H, =CCH₂), 1.74 (quint, J = 7.6, 2 H, =CCH₂CH₂), 1.48–1.18 [m, 10 H, (CH₂)₅CH₃], 0.89 (t, J = 7.0, 3 H, CH₃); ¹³C NMR (75 MHz, CDCl₃): δ = 163.4, 147.8, 143.9, 143.5, 129.1, 126.4, 125.1, 119.5, 114.6, 96.4, 33.2, 31.8, 29.2, 29.1, 28.9, 26.4, 22.6, 14.1; GC/MS = 298 (M⁺, 85), 283 (2), 269 (4), 255 (5), 239 (5), 225 (14), 213 (100), 200 (87), 185 (40), 171 (11), 158 (61), 130 (20); HRMS (ESI-TOF) m/z: [M + H]⁺ Calcd for $C_{18}H_{23}N_2O_2^+$ 299.1754; Found: 299.1757.

3.3.9. 3-Isopentyl-1H-benzo[4,5]imidazo[1,2-c][1,3]oxazin-1-one **2i**

Yield: 210 mg, starting from 312 mg of **1i** (82%) (Table 2, entry 9). Colorless solid, mp: 102–104°C; IR (KBr): ν = 1751 (s), 1667 (m), 1551 (w), 1451 (w), 1366 (s), 1134 (m), 1103 (m), 964 (w), 849 (w), 748 (m) cm⁻¹; ¹H NMR (300 MHz, CDCl₃): δ = 8.21–8.15 (m, 1 H, aromatic), 7.77–7.72 (m, 1 H, aromatic), 7.50–7.37 (m, 2 H, aromatic), 6.48 (s, 1 H, H-4), 2.65–2.55 (m, 2 H, =CCH₂), 1.73–1.56 (m, 3 H, CH₂CH), 0.96 (d, J = 6.2, 6 H, 2 CH₃); ¹³C NMR (75 MHz, CDCl₃): δ = 163.1, 147.4, 144.11, 144.03, 129.3, 126.2, 124.9, 119.7, 114.5, 96.5, 35.3, 31.1, 27.6, 22.3; GC/MS = 256 (M⁺, 100), 241 (6), 227 (2), 214 (10), 200 (56), 185 (25), 171 (5), 158 (61), 143 (4), 130 (14); HRMS (ESI-TOF) m/z: [M + H]⁺ Calcd for $C_{15}H_{17}N_2O_2^+$ 257.1285; Found: 257.1286.

3.3.10. 3-Phenethyl-1H-benzo[4,5]imidazo[1,2-c][1,3]oxazin-1-one **2j**

Yield: 232 mg, starting from 346 mg of **1j** (80%) (Table 2, entry 10). Colorless solid, mp: 159–162 °C; IR (KBr): ν = 1767 (s), 1667 (m), 1558 (w), 1451 (w), 1360 (s), 1103 (m), 988 (m), 864 (m), 756 (s), 694 (s) cm⁻¹; ¹H NMR (300 MHz, CDCl₃): δ = 8.25–8.17 (m, 1 H, aromatic), 7.81–7.72 (m, 1 H, aromatic), 7.53–7.40 (m, 2 H aromatic), 7.35–7.13 (m, 5 H, Ph), 6.45 (s, 1 H, H-4), 3.06 (dist t, J = 7.6, 2 H, CH₂), 2.92 (dist, J = 7.6, 2 H, CH₂); ¹³C NMR (75 MHz, CDCl₃): δ = 161.4, 147.1, 144.0, 143.9, 139.2, 129.3, 128.7, 128.2, 126.7, 126.3, 125.0, 119.8, 114.6, 97.3, 34.8, 32.5; GC/MS = 290 (M⁺, 34), 245 (1), 199 (7), 185 (2), 155 (5), 129 (3), 102 (4), 91 (100); HRMS (ESI-TOF) m/z: [M + H]⁺ Calcd for $C_{18}H_{15}N_2O_2^+$ 291.1128; Found: 291.1126.

3.3.11. 3-(Cyclohexylmethyl)-1H-benzo[4,5]imidazo[1,2-c][1,3]oxazin-1-one **2k**

Yield: 197 mg, starting from 338 mg of **1k** (70%) (Table 2, entry 11). Colorless solid, mp: 135–138°C; IR (KBr): ν = 1767 (s), 1667 (m), 1559 (w), 1451 (w), 1389 (m), 1366 (m), 1096 (w), 964 (w), 748 (m) cm⁻¹; ¹H NMR (500 MHz, CDCl₃) δ = 8.21 (d, J = 7.7, 1 H, aromatic), 7.77 (d, J = 8.1, 1 H, aromatic), 7.52–7.40 (m, 2 H, aromatic), 6.49 (s, 1 H, H-4), 2.48 (d, J = 7.0, 2 H, =CCH₂), 1.93–1.62 (m, 6 H, cyclohexyl), 1.39–0.96 (m, 5 H, cyclohexyl); ¹³C NMR (125 MHz, CDCl₃): δ = 161.6, 147.3, 144.2, 129.4, 126.3, 124.9, 119.8, 114.6, 97.7, 41.0, 35.8, 33.0, 26.2, 26.0; GC/MS: m/z = 282 (M⁺, 67), 200 (100), 156 (24), 129 (5); HRMS (ESI-TOF) m/z: [M + H]⁺ Calcd for $C_{17}H_{19}N_2O_2^+$ 283.1441; Found: 283.1448.

3.3.12. 3-(Cyclohex-1-en-1-yl)-1H-benzo[4,5]imidazo[1,2-c][1,3]oxazin-1-one **2l**

Yield: 176 mg, starting from 322 mg of **1l** (66%) (Table 2, entry 12). Colorless solid, mp: 191–195 °C; IR (KBr): ν = 1767 (s), 1636 (m), 1420 (w), 1366 (m), 1281 (w), 1180 (w), 1111 (m), 1026 (w), 833 (w), 748 (m) cm⁻¹; ¹H NMR (300 MHz, CDCl₃) δ = 8.24–8.16 (m, 1 H, aromatic), 7.81–7.71 (m, 1 H, aromatic), 7.53–7.39 (m, 2 H, aromatic), 7.00–6.90 (m, 1 H, =CH), 6.55 (s, 1 H, H-4), 2.40–2.24 (m, 4 H, cyclohexenyl), 1.86–1.74 (m, 2 H, cyclohexenyl), 1.74–1.62 (m, 2 H, cyclohexenyl); ¹³C NMR (75 MHz, CDCl₃): δ = 157.7, 148.1, 144.5, 143.6, 134.0, 129.5, 127.1, 126.2, 124.9, 119.6, 114.5, 92.8, 25.9, 23.9, 22.0, 21.5; GC/MS = 266

(M$^+$, 100), 237 (7), 221 (23), 185 (26), 157 (9); HRMS (ESI-TOF) m/z: [M + H]$^+$ Calcd for C$_{16}$H$_{15}$N$_2$O$_2{}^+$ 267.1128; Found: 267.1129.

3.3.13. 3-(Methoxymethyl)-1H-benzo[4,5]imidazo[1,2-c][1,3]oxazin-1-one **2m**

Yield: 138 mg, starting from 286 mg of **1m** (60%) (Table 2, entry 13). Yellow solid, mp: 122–125°C; IR (KBr): ν = 1751 (s), 1667 (m), 1558 (m), 1443 (m), 1381 (s), 1173 (m), 1103 (s), 957 (w), 748 (s) cm^{-1}; ^{1}H NMR (500 MHz, CDCl$_3$) δ = 8.23–8.18 (m, 1 H, aromatic), 7.82–7.75 (m, 1 H, aromatic), 7.53–7.43 (m, 2 H, aromatic), 6.81–6.78 (m, 1 H, H-4), 4.35 (s, 2 H, CH$_2$OCH$_3$), 3.53 (s, 3 H, OCH$_3$); ^{13}C NMR (125 MHz, CDCl$_3$): δ = 158.2, 146.8, 144.1, 143.5, 129.4, 126.4, 125.3, 120.0, 114.6, 97.2, 69.6, 59.4; GC/MS: m/z = 230 (M$^+$, 89), 199 (5), 185 (100), 171 (10), 157 (48), 129 (8); HRMS (ESI-TOF) m/z: [M + H]$^+$ Calcd for C$_{12}$H$_{11}$N$_2$O$_3{}^+$ 231.0764; Found: 231.0768.

3.3.14. Methyl 3-(1-oxo-1H-benzo[4,5]imidazo[1,2-c][1,3]oxazin-3-yl)propanoate **2n**

Yield: 201 mg, starting from 328 mg of **1n** (74%) (Table 2, entry 14). Colorless solid, mp: 189–193°C; IR (KBr): ν = 1767 (s), 1736 (s), 1667 (m), 1435 (w), 1366 (w), 1173 (m), 996 (m), 895 (w), 841 (w), 772 (m) cm^{-1}; ^{1}H NMR (500 MHz, CDCl$_3$) δ = 8.19 (d, J = 7.7, 1 H, aromatic), 7.77 (d, J = 8.1, 1 H, aromatic), 7.51–7.41 (m, 2 H, aromatic), 6.58 (s, 1 H, H-4), 3.72 (s, 3 H, CO$_2$CH$_3$), 2.97 (t, J = 7.2, 2 H, =CCH$_2$), 2.79 (t, J = 7.2, 2 H, CH$_2$CO$_2$CH$_3$); ^{13}C NMR (125 MHz, CDCl$_3$): δ = 171.8, 160.4, 147.0, 144.1, 143.7, 126.4, 125.2, 119.9, 114.6, 97.5, 52.1, 30.5, 28.4; GC/MS: m/z = 272 (M$^+$, 61), 243 (15), 212 (100), 199 (35), 185 (33), 169 (20), 157 (35); HRMS (ESI-TOF) m/z: [M + H]$^+$ Calcd for C$_{14}$H$_{13}$N$_2$O$_4{}^+$ 273.0870; Found: 273.0874.

3.3.15. 3-(2-(tert-Butoxy)ethyl)-1H-benzo[4,5]imidazo[1,2-c][1,3]oxazin-1-one **2o'**

Yield: 189 mg, starting from 286 mg of **1o** (66%) (Table 2, entry 15). Colorless solid, mp: 189–193°C; IR (KBr): ν = 1774 (s), 1666 (m), 1551 (w), 1389 (w), 1366 (w), 1204 (w), 1111 (w), 1080 (m), 756 (m) cm^{-1}; ^{1}H NMR (500 MHz, CDCl$_3$) δ = 8.25–8.21 (m, 1 H, aromatic), 7.82–7.77 (m, 1 H, aromatic), 7.52–7.42 (m, 2 H, aromatic), 6.65 (dist t, J = 0.8, 1 H, H-4), 3.73 (t, J = 6.1, 2 H, CH$_2$Ot-Bu), 2.83 (td, J = 6.1, 0.8, 2 H, =CCH$_2$), 1.20 (s, 9 H,); ^{13}C NMR (125 MHz, CDCl$_3$): δ = 160.5, 147.4, 144.2, 144.0, 129.4, 126.3, 125.0, 119.8, 114.6, 98.0, 73.5, 57.7, 34.5, 27.4; GC/MS: m/z = 286 (M$^+$, 21), 213 (12), 200 (100), 171 (16), 156 (22); HRMS (ESI-TOF) m/z: [M + H]$^+$ Calcd for C$_{16}$H$_{19}$N$_2$O$_3{}^+$ 287.1390; Found: 287.1395.

3.3.16. 2-(Hex-1-yn-1-yl)-6-nitro-1H-benzo[d]imidazole **3f**

Yield: 49 mg, starting from 343 mg of **1f** (20%) (Table 2, entry 6). Colorless solid, mp: 138–140 °C; IR (KBr): ν = 2230 (w), 1520 (s), 1474 (w), 1435 (w), 1343 (s), 1065 (w), 818 (m) cm^{-1}; ^{1}H NMR (500 MHz, DMSO-d_6): δ = 8.41 (s, br, 1 H, H-3), 8.14 (dd, J = 8.9, 2.2, 1 H, H-5), 7.69 (d, J = 8.9, 1 H, H-4), 2.58 (t, J = 7.2, 2 H, ≡CCH$_2$), 1.60 (quint, J = 7.2, 2 H, CH$_2$CH$_2$CH$_3$), 1.49 (hexuplet, J = 7.2, 2 H, CH$_2$CH$_3$), 0.95 (t, J = 7.2, 3 H, CH$_3$) (Note: the NH signal was incorporated into the broad HOD signal at 3.49 ppm); ^{13}C NMR (125 MHz, DMSO-d_6): δ = 143.1, 139.6, 118.3, 114.3 (br), 95.8, 71.6, 29.5, 21.4, 18.1, 13.3; GC/MS: m/z = 243 (M$^+$, 100), 228 (48), 214 (73), 201 (93), 182 (41), 168 (54), 155 (57), 127 (27); HRMS (ESI-TOF) m/z: [M + H]$^+$ Calcd for C$_{13}$H$_{14}$N$_3$O$_2{}^+$ 244.1081; Found: 244.1081.

3.3.17. 2-(Hex-1-yn-1-yl)-5-nitro-1H-benzo[d]imidazole **3g**

Yield: 75 mg, starting from 343 mg of **1g** (31%) (Table 2, entry 7). Yellow solid, mp: 145–148 °C; IR (KBr): ν = 2237 (w), 1520 (s), 1474 (w), 1435 (w), 1366 (w), 1342 (s), 1065 (m), 818 (m), 741 (m) cm^{-1}; ^{1}H NMR (500 MHz, CDCl$_3$): δ = 8.75 (s, 1 H, H-4), 8.29 (d, J = 8.8, 1 H, H-6), 7.83 (d, J = 8.8, 1 H, H-7), 2.48 (t, J = 7.3, 2 H, ≡CCH$_2$), 1.50 (quint, J = 7.3, 2 H, CH$_2$CH$_2$CH$_3$), 1.33 (hexuplet, J = 7.3, 2 H, CH$_2$CH$_3$), 0.78 (t, J = 7.3, 3 H, CH$_3$) (Note: the NH signal was too broad to be detected); ^{13}C NMR (125 MHz, CDCl$_3$): δ = 144.4, 140.4, 119.2, 115.1 (br), 112.6 (br), 98.2, 71.0, 29.9, 22.0, 19.1, 13.4; GC/MS: m/z = 243 (M$^+$, 100), 228 (44), 214 (71), 201 (96), 182 (40), 168 (54), 155 (56); HRMS (ESI-TOF) m/z: [M + H]$^+$ Calcd for C$_{13}$H$_{14}$N$_3$O$_2{}^+$ 244.1081; Found: 244.1082.

4. Conclusions

In conclusion, we have reported that simple and inexpensive $ZnCl_2$ is able to promote the heterocyclization of N-Boc-2-alkynylbenzimidazoles under mild conditions (40 °C in CH_2Cl_2 for 3h), giving access to new polycyclic heterocycles, 1H-benzo[4,5]imidazo[1,2-c][1,3]oxazin-1-ones. While in the previous literature $ZnCl_2$ was reported to promote complete N-Boc deprotection with elimination of isobutene and CO_2, in the present process it assisted the 6-endo-dig heterocyclization of the carbamate intermediate with incorporation of the carbamate group into the final polyheterocyclic derivative. $ZnCl_2$ thus played a dual role, by promoting the Boc deprotection of the substrate with elimination of the tert-butyl carbonation (which could be trapped by substrates bearing a nucleophilic group) and activating the triple bond toward the intramolecular nucleophilic attack by the carbamate moiety. The benzimidazoxazinone derivatives have been obtained in moderate to high yields starting from differently substituted substrates, and the structure of representative products has been confirmed by X-ray diffraction analysis.

Supplementary Materials: The following are available online. Preparation and characterization of N-Boc-2-alkynylbenzimidazole substrates **1a–1o**, X-ray crystallographic data for products **2a**, **2c**, and **2f**, Copies of HRMS, ^{1}H NMR, and ^{13}CNMR spectra.

Author Contributions: Conceptualization: B.G. and L.V.; methodology: L.V., R.A., M.P., C.C., M.A.C., P.V., B.G.; validation: R.A., M.P., and L.V.; investigation: L.V., R.A., M.P., C.C., M.A.C., P.V.; writing—original draft preparation: B.G.; writing—review and editing: B.G.; supervision: B.G.; funding acquisition, B.G. All authors have read and agreed to the published version of the manuscript.

Funding: Financial support by MIUR PRIN 2017YJMPZN project (Mussel-inspired functional biopolymers for underwater adhesion, surface/interface derivatization and nanostructure/composite self-assembly-MUSSEL) to B.G. is acknowledged.

Institutional Review Board Statement: Not applicable.

Informed Consent Statement: Not applicable.

Data Availability Statement: The data presented in this study are available on request from the corresponding authors.

Conflicts of Interest: The authors declare no conflict of interest.

Sample Availability: Samples of the compounds **2a–2n**, **2o'**, **3f**, and **3g** are available from the authors.

References

1. Zheng, L.; Hua, R. Recent Advances in Construction of Polycyclic Natural Product Scaffolds via One-Pot Reactions Involving Alkyne Annulation. *Front. Chem.* **2020**, *8*, 580355. [CrossRef]
2. Hong, F.-L.; Ye, L.-W. Transition Metal-Catalyzed Tandem Reactions of Ynamides for Divergent N-Heterocycle Synthesis. *Acc. Chem. Res.* **2020**, *53*, 2003–2019. [CrossRef] [PubMed]
3. Kaur, N.; Verma, Y.; Grewal, P.; Ahlawat, N.; Bhardwaj, P.; Jangid, N.K. Palladium acetate assisted synthesis of five-membered N-polyheterocycles. *Synth. Commun.* **2020**, *50*, 1567–1621. [CrossRef]
4. Gabriele, B.; Mancuso, R.; Veltri, L.; Ziccarelli, I.; Della Ca', N. Palladium-Catalyzed Double Cyclization Processes Leading to Polycyclic Heterocycles: Recent Advances. *Eur. J. Org. Chem.* **2019**, *2019*, 5073–5092. [CrossRef]
5. Wang, R.; Xie, X.; Liu, H.; Zhou, Y. Rh(III)-Catalyzed C–H Bond Activation for the Construction of Heterocycles with sp^3-Carbon Centers. *Catalysts* **2019**, *9*, 823. [CrossRef]
6. Ghosh, A.; Carter, R.G. Recent Syntheses and Strategies toward Polycyclic Gelsemium Alkaloids. *Angew. Chem. Int. Ed.* **2019**, *58*, 681–694. [CrossRef]
7. Passador, K.; Thorimbert, S.; Botuha, C. Heteroaromatic Rings of the Future': Exploration of Unconquered Chemical Space. *Synthesis* **2019**, *51*, 384–398.
8. Hyland, I.K.; O'Toole, R.F.; Smith, J.A.; Bissember, A.C. Progress in the Development of Platelet-Activating Factor Receptor (PAFr) Antagonists and Applications in the Treatment of Inflammatory Diseases. *ChemMedChem* **2018**, *13*, 1873–1884. [CrossRef]
9. Hemmerling, F.; Hahn, F. Biosynthesis of oxygen and nitrogen-containing heterocycles in polyketides. *Beilstein J. Org. Chem.* **2016**, *12*, 1512–1550. [CrossRef]
10. Shokova, E.A.; Kovalev, V.V. Biological Activity of Adamantane-Containing Mono- and Polycyclic Pyrimidine Derivatives (A Review). *Pharm. Chem. J.* **2016**, *50*, 63–75. [CrossRef]

11. Fizer, M.; Slivka, M. Synthesis of [1,2,4]triazolo[1,5-*a*]pyrimidine (microreview). *Chem. Heterocycl. Compds.* **2016**, *52*, 155–157. [CrossRef]
12. Li, J.; Yang, F.; Hu, W.; Ren, B.; Chen, Z.-S.; Ji, K. Gold(I)-catalyzed tandem cyclization of cyclopropylidene-tethered propargylic alcohols: An approach to functionalized naphtho[2,3-*c*]pyrans. *Chem. Commun.* **2020**, *56*, 9154–9157. [CrossRef] [PubMed]
13. He, Y.; Li, Z.; Robyens, K.; Van Meervelt, L.; Van der Eycken, R.V. A Gold-Catalyzed Domino Cyclization Enabling Rapid Construction of Diverse Polyheterocyclic Frameworks. *Angew. Chem. Int. Ed.* **2018**, *57*, 272–276. [CrossRef]
14. Alcaide, B.; Almendros, P.; Fernández, I.; Herrea, F.; Luna, A. Gold-Catalyzed Divergent Ring-Closing Modes of Indole-Tethered Amino Allenynes. *Chem. Eur. J.* **2018**, *24*, 1448–1454. [CrossRef]
15. Li, Z.; Song, L.; Van Meervelt, L.; Tian, G.; Van der Eycken, E.V. Cationic Gold(I)-Catalyzed Cascade Bicyclizations for Divergent Synthesis of (Spiro)polyheterocycles. *ACS Catal.* **2018**, *8*, 6388–6393. [CrossRef]
16. Zhang, J.-h.; Wei, Y.; Shi, M. Gold-catalyzed ring enlargement and cycloisomerization of alkynylamide tethered alkylidenecyclopropanes. *Org. Chem. Front.* **2018**, *5*, 2980–2985. [CrossRef]
17. Ito, M.; Kawasaki, R.; Kanyiva, K.S.; Shibata, T. Construction of a Polycyclic Conjugated System Containing a Dibenzazepine Moiety by Cationic Gold(I)-Catalyzed Cycloisomerization. *Eur. J. Org. Chem.* **2016**, *2016*, 5234–5237. [CrossRef]
18. Kumar, R.; Arigela, R.K.; Samala, S.; Kundu, B. Diversity Oriented Synthesis of Indoloazepinobenzimidazole and Benzimidazotriazolobenzodiazepine from N^1-Alkyne-1,2-diamines. *Chem. Eur. J.* **2015**, *21*, 18828–18833. [CrossRef]
19. Chen, M.; Sun, N.; Xu, W.; Zhao, J.; Wang, G.; Liu, Y. Gold-Catalyzed Ring Expansion of Alkynyl Heterocycles through 1,2-Migration of an Endocyclic Carbon–Heteroatom Bond. *Chem. Eur. J.* **2015**, *21*, 18571–18575. [CrossRef] [PubMed]
20. Zheng, Y.; Bao, M.; Yao, R.; Qiu, L.; Xu, X. Palladium-catalyzed carbene/alkyne metathesis with enynones as carbene precursors: Synthesis of fused polyheterocycles. *Chem. Commun.* **2018**, *54*, 350–353. [CrossRef] [PubMed]
21. Feng, Y.; Tian, N.; Li, Y.; Jia, C.; Li, X.; Wang, L.; Ciu, X. Construction of Fused Polyheterocycles through Sequential [4 + 2] and [3 + 2] Cycloadditions. *Org. Lett.* **2017**, *19*, 1658–1661. [CrossRef]
22. Kumar, S.; Saunthwal, R.K.; Aggarwal, T.; Kotla, S.K.R.; Verma, A.K. Palladium meets copper: One-pot tandem synthesis of pyrido fused heterocycles *via* Sonogashira conjoined electrophilic cyclization. *Org. Biomol. Chem.* **2016**, *14*, 9063–9071. [CrossRef] [PubMed]
23. Dethe, D.H.; Boda, R. A Novel Pd-Catalysed Annulation Reaction for the Syntheses of Pyrroloindoles and Pyrroloquinolines. *Chem. Eur. J.* **2016**, *22*, 106–110. [CrossRef] [PubMed]
24. Wu, X.; Ji, H. Rhodium-Catalyzed [4 + 1] Cyclization via C–H Activation for the Synthesis of Divergent Heterocycles Bearing a Quaternary Carbon. *J. Org. Chem.* **2018**, *83*, 4650–4656. [CrossRef] [PubMed]
25. Youn, S.W.; Yoo, H.J. One-Pot Sequential N-Heterocyclic Carbene/Rhodium(III) Catalysis: Synthesis of Fused Polycyclic Isocoumarins. *Adv. Synth. Catal.* **2017**, *359*, 2176–2183. [CrossRef]
26. Ghorai, D.; Choudhury, J. Rhodium(III)–N-Heterocyclic Carbene-Driven Cascade C–H Activation Catalysis. *ACS Catal.* **2015**, *5*, 2692–2696. [CrossRef]
27. Kozak, J.A.; Dodd, J.M.; Harrison, T.J.; Jardine, K.J.; Patrick, B.O.; Dake, G.R. Enamides and Enesulfonamides as Nucleophiles: Formation of Complex Ring Systems through a Platinum(II)-Catalyzed Addition/Friedel−Crafts Pathway. *J. Org. Chem.* **2009**, *74*, 6929–6935. [CrossRef]
28. Marion, F.; Coulomb, J.; Servais, A.; Courillon, C.; Fensterbank, L.; Malacria, M. Radical cascade cyclizations and platinum(II)-catalyzed cycloisomerizations of ynamides. *Tetrahedron* **2006**, *62*, 3856–3871. [CrossRef]
29. Mamane, V.; Hannen, P.; Fürstner, A. Synthesis of Phenanthrenes and Polycyclic Heteroarenes by Transition-Metal Catalyzed Cycloisomerization Reactions. *Chem. Eur. J.* **2004**, *10*, 4556–4575. [CrossRef]
30. Ghosh, K.; Shankar, R.; Rit, R.K.; Dubey, G.; Bharatam, P.V.; Sahoo, A.K. Sulfoximine-Assisted One-Pot Unsymmetrical Multiple Annulation of Arenes: A Combined Experimental and Computational Study. *J. Org. Chem.* **2018**, *83*, 9667–9681. [CrossRef]
31. Miclo, Y.; Garcia, P.; Evanno, Y.; George, P.; Sevrin, M.; Malacria, M.; Gandon, V.; Aubert, C. Synthesis of Orthogonally Protected Angular Nitrogen Polyheterocycles via CpCo-Catalyzed Pyridine Formation. *Synlett* **2010**, *2010*, 314–2318.
32. Hoshimoto, Y.; Ashida, K.; Sasaoka, Y.; Kumar, R.; Kamikawa, K.; Verdaguer, X.; Riera, A.; Ohashi, M.; Ogoshi, S. Efficient Synthesis of Polycyclic γ-Lactams by Catalytic Carbonylation of Ene-Imines via Nickelacycle Intermediates. *Angew. Chem. Int. Ed.* **2017**, *56*, 8206–8210. [CrossRef] [PubMed]
33. Liu, X.; Wang, Z.-S.; Zhai, T.-Y.; Luo, C.; Zhang, Y.-P.; Chen, Y.-B.; Deng, C.; Liu, R.-S.; Ye, L.-W. Copper-Catalyzed Azide–Ynamide Cyclization to Generate α-Imino Copper Carbenes: Divergent and Enantioselective Access to Polycyclic N-Heterocycles. *Angew. Chem. Int. Ed.* **2020**, *59*, 17984–17990. [CrossRef] [PubMed]
34. Mao, X.-F.; Zhu, X.-P.; Li, D.-Y.; Liu, P.-N. Cu-Catalyzed Cascade Annulation of Alkynols with 2-Azidobenzaldehydes: Access to 6*H*-Isochromeno[4,3-*c*]quinolone. *J. Org. Chem.* **2017**, *82*, 7032–7039. [CrossRef] [PubMed]
35. Ho, H.E.; Oniwa, K.; Yamamoto, Y.; Jin, T. N-Methyl Transfer Induced Copper-Mediated Oxidative Diamination of Alkynes. *Org. Lett.* **2016**, *18*, 2487–2490. [CrossRef] [PubMed]
36. Mandadapu, A.K.; Sharma, S.K.; Gupta, S.; Krishna, D.G.V.; Kundu, B. Unprecedented Cu-Catalyzed Coupling of Internal 1,3-Diynes with Azides: One-Pot Tandem Cyclizations Involving 1,3-Dipolar Cycloaddition and Carbocyclization Furnishing Naphthotriazoles. *Org. Lett.* **2011**, *13*, 3162–3165. [CrossRef] [PubMed]
37. Bakholdina, A.; Lukin, A.; Bakulina, O.; Guranova, N.; Krasavin, M. Dual use of propargylamine building blocks in the construction of polyheterocyclic scaffolds. *Tetrahedron Lett.* **2020**, *61*, 151970. [CrossRef]

38. Habert, L.; Sallio, R.; Durandetti, M.; Gosmini, C.; Gillaizeau, I. Zinc Chloride Mediated Synthesis of 3*H*-Oxazol-2-one and Pyrrolo-oxazin-1-one from Ynamide. *Eur. J. Org. Chem.* **2019**, *2019*, 5175–5179. [CrossRef]
39. Muralidhar, B.; Reddy, S.R. Zn(II) Chloride Promoted Benzannulation Strategy for One-Pot Regioselective Synthesis of 6*H*-Benzo[*c*]chromenes. *ChemistrySelect* **2017**, *2*, 2539–2543. [CrossRef]
40. Li, L.; Zhou, B.; Wang, Y.-H.; Shu, C.; Pan, Y.-F.; Lu, X.; Ye, L.-W. Zinc-Catalyzed Alkyne Oxidation/C-H Functionalization: Highly Site-Selective Synthesis of Versatile Isoquinolones and β-Carbolines. *Angew. Chem. Int. Ed.* **2015**, *54*, 8245–8249. [CrossRef]
41. Kim, H.; Tung, T.T.; Park, S.B. Privileged Substructure-Based Diversity-Oriented Synthesis Pathway for Diverse Pyrimidine-Embedded Polyheterocycles. *Org. Lett.* **2013**, *15*, 5814–5817. [CrossRef]
42. Liu, Y.; Zhen, W.; Dai, W.; Wang, F.; Li, X. Silver(I)-Catalyzed Addition-Cyclization of Alkyne-Functionalized Azomethines. *Org. Lett.* **2013**, *15*, 874–877. [CrossRef]
43. Yamamoto, Y.; Nishiyama, M.; Imagawa, H.; Nishizawa, M. Hg(OTf)$_2$-Catalyzed cyclization of alkynyl *tert*-butylcarbonate leading to cyclic enol carbonate. *Tetrahedron Lett.* **2006**, *47*, 8369–8373. [CrossRef]
44. Krishnan, K.K.; Ujwaldev, S.M.; Saranya, S.; Anilkumar, G.; Beller, M. Recent Advances and Perspectives in the Synthesis of Heterocycles via Zinc Catalysis. *Adv. Synth. Catal.* **2019**, *361*, 382–404. [CrossRef]
45. Saranya, S.; Harry, N.A.; Ujwaldev, S.M.; Anilkumar, G. Recent Advances and Perspectives on the Zinc-Catalyzed Nitroaldol (Henry) Reaction. *Asian J. Org. Chem.* **2017**, *6*, 1349–1360. [CrossRef]
46. Thankachan, A.P.; Asha, S.; Sindhu, K.S.; Anilkumar, G. An overview of Zn-catalyzed enantioselective aldol type C-C bond formation. *RSC Adv.* **2015**, *5*, 62179–62193. [CrossRef]
47. Wu, X.-F. Non-Redox-Metal-Catalyzed Redox Reactions: Zinc Catalysts. *Chem. Asian J.* **2012**, *7*, 2502–2509. [CrossRef]
48. Wu, X.-F.; Neumann, H. Zinc-Catalyzed Organic Synthesis: C-C, C-N, C-O Bond Formation Reactions. *Adv. Synth. Catal.* **2012**, *354*, 3141–3160. [CrossRef]
49. Wutts, P.G.M. *Greene's Protective Groups in Organic Synthesis*, 5th ed.; John Wiley & Sons, Inc.: Hoboken, NJ, USA, 2014; pp. 755–939.
50. Langille, E.; Bottaro, C.S.; Drouin, A. A novel use of catalytic zinc-hydroxyapatite columns for the selective deprotection of *N*-tert-butyloxycarbonyl (BOC) protecting group using flow chemistry. *J. Flow Chem.* **2020**, *10*, 377–387. [CrossRef]
51. Vu, H.-D.; Renault, J.; Roisnel, T.; Robert, C.; Jéhan, P.; Gouault, N.; Uriac, P. Reactivity of *N*-Boc-Protected Amino-Ynones in the Presence of Zinc Chloride: Formation of Acetylenic Cyclic Imines and Their Palladium Complexes. *Eur. J. Org. Chem.* **2015**, *2015*, 4868–4875. [CrossRef]
52. Nigam, S.C.; Mann, A.; Taddei, M.; Wermuth, C.-G. Selective Removal of the Tert-Butoxycarbonyl Group from Secondary Amines: ZnBr$_2$ as the Deprotecting Reagent. *Synth. Commun.* **1989**, *19*, 3139–3142. [CrossRef]
53. Wu, Y.-Q.; Limburg, D.C.; Wilkinson, D.E.; Vaal, M.J.; Hamilton, G.S. A mild deprotection procedure for *tert*-butyl esters and *tert*-butyl ethers using ZnBr$_2$ in methylene chloride. *Tetrahedron Lett.* **2000**, *41*, 2847–2849. [CrossRef]
54. Kaul, R.; Brouillette, Y.; Sajjadi, Z.; Hansford, K.A.; Lubell, W.D. Selective *tert*-Butyl Ester Deprotection in the Presence of Acid Labile Protecting Groups with Use of ZnBr$_2$. *J. Org. Chem.* **2004**, *69*, 6131–6133. [CrossRef] [PubMed]

Communication

Synthesis of Functionalized Indoles via Palladium-Catalyzed Cyclization of *N*-(2-allylphenyl) Benzamide: A Method for Synthesis of Indomethacin Precursor

Zhe Chang, Tong Ma, Yu Zhang, Zheng Dong, Heng Zhao and Depeng Zhao *

School of Pharmaceutical Sciences, Sun Yat-sen University, Guangzhou 510006, China; changzh5@mail2.sysu.edu.cn (Z.C.); mt11072020@126.com (T.M.); zhangy637@mail2.sysu.edu.cn (Y.Z.); dongzh6@mail2.sysu.edu.cn (Z.D.); zhaoh5@mail2.sysu.edu.cn (H.Z.)
* Correspondence: zhaodp5@mail.sysu.edu.cn

Academic Editor: Bartolo Gabriele
Received: 20 February 2020; Accepted: 7 March 2020; Published: 9 March 2020

Abstract: We developed an efficient method for synthesis of substituted *N*-benzoylindole via Pd(II)-catalyzed C–H functionalization of substituted *N*-(2-allylphenyl)benzamide. The reaction showed a broad substrate scope (including *N*-acetyl and *N*-Ts substrates) and substituted indoles were obtained in good to excellent yields. The most distinctive feature of this method lies in the high selectivity for *N*-benzoylindole over benzoxazine, and this is the first example of Pd(II)-catalyzed synthesis of substituted *N*-benzoylindole. Notably, this new method was applied for the synthesis of key intermediate of indomethacin.

Keywords: palladium; indole; indomethacin; C-H functionalization

1. Introduction

Indole skeletons are one of the most valuable heterocycles, due to their diverse biological activities and broad applications in functionalized materials and chemistry [1–3]. Substituted indoles exist extensively in nature and pharmaceuticals (Figure 1) [2,4,5]. As a result, many methods have been developed for the synthesis of indoles, including pioneering studies by Fischer, [6] Larock [7,8], Buchwald [9–11], and Hegedus [12–15]. Recently, an increasing number of approaches for the synthesis of indoles by employing transition-metal-catalyzed oxidative C–H bond functionalization has been reported (Figure 2a) [16–26]. These methods showed significant improvement with regard to the substrate scope and reaction conditions [27–31], but *N*-substituents in these reports are restricted to H, acetyl, Ts (tosyl), and Ms (mesyl). In addition, Rh(III)-catalyzed tandem C-H allylation and oxidative cyclization of anilides with allyl carbonates or acetates have been developed, albeit with costly Rh complexes [32–34]. Despite these achievements, it is still of great value to develop methods for synthesizing substituted indoles, especially *N*-benzoyl with low cost and wide variety of substituents. Substituted *N*-benzoylindole is one of the most attractive skeletons, since it is a privileged structure of many pharmacologically active compounds such as indomethacin. Besides, the only method reported to construct substituted *N*-benzoyl indole is the C-H-amination of styrenes using hypervalent iodine as the oxidant (Figure 2b) [35,36]. However, it should be noted that the oxidants of these two methods are not commercially available. A direct approach employing simple and readily available catalyst and oxidant remains a challenge for the synthesis of substituted *N*-benzoylindole. In this context, our strategy is to use commercially available and inexpensive catalyst Pd(OAc)$_2$ and oxidant benzoquinone (BQ) for C–H functionalization to construct substituted *N*-benzoyl indoles (Figure 2c). It was reported that either aminopalladated or π-allyl Pd intermediates would be generated in palladium-catalyzed allylic C–H oxidation reaction with the usage of ambident O/N nucleophiles [37–39]. Our method can avoid the generation of the π-allyl Pd intermediates and obtain corresponding N-benzoyl indoles. More importantly, the synthetic utility of this method is further demonstrated by the synthesis of essential skeleton of indomethacin.

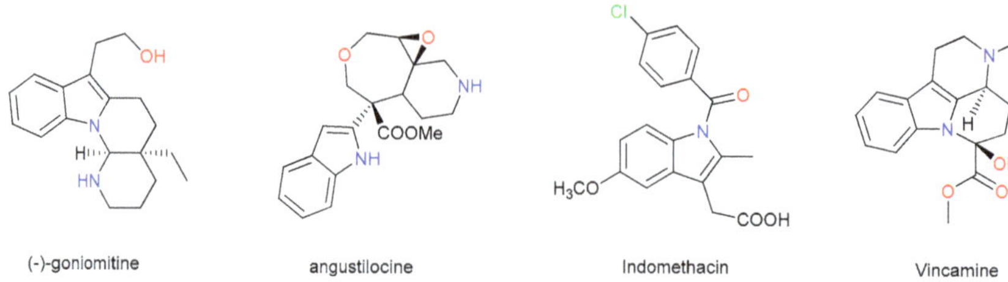

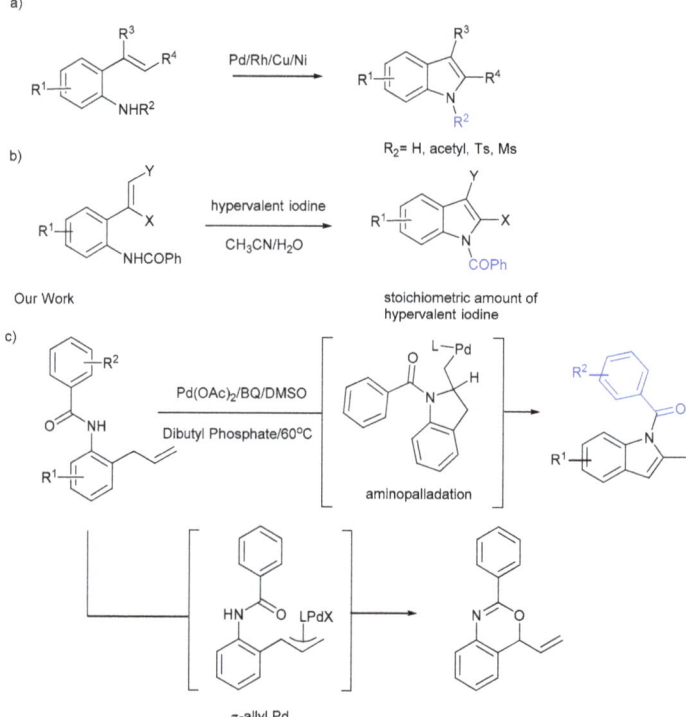

Figure 1. Representative molecules based on indole skeleton.

Figure 2. Methods for synthesis of substituted N-benzoylindole. (a) The synthesis of indoles by employing transition-metal-catalyzed. (b) The synthesis of N-benzoylindole by hypervalent iodine as the oxidant. (c) This work.

2. Results and Discussion

We began our study by the reaction of **1a** in the presence of 10 mol% of Pd (OAc)$_2$ as the catalyst and BQ (1.5 equiv.) as the oxidant in MeCN at room temperature (Table 1, Entry 1). Gratifyingly, the desired product **2a** was obtained in 8% yield along with a by-product benzoxazine **3** formed via allylic C-H cleavage (**2a**/**3** = 1/1) [38,39]. Encouraged by this result, we further systematically optimized the reaction conditions to improve the conversion of the reaction and inhibit the formation of benzoxazine 3. When the reaction was performed at elevated temperature of 60 °C, a slightly higher yield of **2a** was achieved and the ratio of **2a** to **3** was also enhanced to 4:1 (Table 1, Entry 3). Interestingly, addition of a stoichiometric amount of AcOH facilitated this reaction to give a higher selectivity (Table 1, Entry 4), improving the ratio to 10:1. Encouraged by this result, we evaluated several acids as additives of the reaction, as shown in Table 1 (Entries 5–8). We were excited to find that using dibutyl phosphate (DBP) as the acid led to a significantly higher yield and the ratio of **2a** to **3** was also improved to more than 20:1 (71% yield, Table 1, Entry 8). With dibutyl phosphate as the optimal additive, two other Pd catalysts were tested, but no satisfactory results were obtained (Table 1, Entries 9 and 10). Next, a survey of other solvents was then carried out (Table 1, Entries 11–14). To our delight, the yield could be further increased to 77% by using DMSO as the solvent (Table 1, Entry 11). Eventually, when 2 equiv. of BQ was used, the reaction gave the desired product in excellent yield (81%, Table 1, Entry 15).

Table 1. Optimization of the Reaction Conditions.[a]

Entry	Catalyst	Additive	Solvent	T [°C]	Yield [b] (2a:3) [c]
1	Pd(OAc)$_2$	-	CH$_3$CN	RT	8 (1:1)
2	Pd(OAc)$_2$	-	CH$_3$CN	45	26 (2:1)
3	Pd(OAc)$_2$	-	CH$_3$CN	60	33 (4:1)
4	Pd(OAc)$_2$	AcOH	CH$_3$CN	60	12 (10:1)
5	Pd(OAc)$_2$	PhCOOH	CH$_3$CN	60	22 (10:1)
6	Pd(OAc)$_2$	TFA	CH$_3$CN	60	26 (9:1)
7	Pd(OAc)$_2$	Ph$_2$PO$_2$H	CH$_3$CN	60	36 (15:1)
8	Pd(OAc)$_2$	(BuO)$_2$PO$_2$H	CH$_3$CN	60	71 (>20:1)
9	PdCl$_2$	(BuO)$_2$PO$_2$H	CH$_3$CN	60	23 (>20:1)
10	White catalyst [d]	(BuO)$_2$PO$_2$H	CH$_3$CN	60	9 (>20:1)
11	Pd(OAc)$_2$	(BuO)$_2$PO$_2$H	DMSO	60	77 (>20:1)
12	Pd(OAc)$_2$	(BuO)$_2$PO$_2$H	dioxane	60	52 (>20:1)
13	Pd(OAc)$_2$	(BuO)$_2$PO$_2$H	THF	60	45 (>20:1)
14	Pd(OAc)$_2$	(BuO)$_2$PO$_2$H	DMF	60	36 (>20:1)
15	**Pd(OAc)$_2$**	**(BuO)$_2$PO$_2$H**	**DMSO**	**60**	**81 (>20:1)**

[a] Reaction conditions: **1a** (0.2 mmol), Pd(II) catalyst (10 mol%), additive (1.5 equiv.), BQ (Entries 1–14,1.5 equiv.; Entry 15, 2.0 equiv.), solvent (2.0 mL), 24 h; [b] Isolated yield of **2a**; [c] Determined by ^{1}H NMR analysis of the crude residue. [d] 1,2-Bis(phenylsulfinyl)ethane palladium(II) acetate.

With the optimized reaction conditions in hand, we turned to explore the substrate scope of this reaction. The reactions of *N*-(2-allylphenyl) benzamides with substituent (**R**) at the positions of benzamide aryl group were initially examined. As shown in Table 2, all substrates proceeded smoothly to afford the corresponding indole in moderate to good yields (62–90%). In general, better yields were found for substrates with electron-rich (**2g, 2h, 2i,** and **2u**) rather than electron-poor anilides (**2b, 2c, 2d, 2e,** and **2f**). Prolonged reaction times were required for substrates with the latter substituents (**2b, 2c,** and **2d**). Substrates **2b, 2c,** and **2d** with Cl substituent at the *meta-*, *ortho-*, and *para-* position of the benzamide aryl group, respectivelu, were also studied. The results indicate that a relatively lower yield was observed for **2b** with *ortho*-Cl comparing with **2c** and **2d**. Additionally, products with substituents at the *meta*-position (**2d, 2e,** and **2i**) can be obtained in higher yields than those with *para*-substituents

(**2c**, **2f**, and **2h**). Similarly, indoles with two substituents on the phenyl ring (**2j** and **2u**) were also obtained in pretty good yield.

Table 2. Substrate scope of substrates with substituents at the positions (R) [a,b] Conditions: See Supplementary Materials for details.

Reaction scheme: 1b-j → 2b-j with Pd(OAc)$_2$/BQ/DMSO, Dibutyl Phosphate/60 °C	

2b	2c	2d	2e	2f
48 h, yield 62%	48 h, yield 73%	48 h, yield 65%	24 h, yield 73%	24 h, yield 79%

2g	2h	2i	2j	2u
24 h, yield 81%	24 h, yield 90%	24 h, yield 84%	24 h, yield 76%	24 h, yield 73%

[a] All reactions were carried out in 0.2 mmol scale. [b] Yields referred to here are isolated yields.

Subsequently, we investigated the effects of substituents (R^1) residing on the aromatic moiety of the N-(2-allylphenyl) benzamide (Table 3). All the substrates **1k–t** gave the desired products **2k–t** in satisfactory yields (67–78%), but 48 h were required for the starting materials to be consumed completely in most cases. The reaction yields were not significantly influenced by the electronic properties of R^1 substituent. The effect of the same substituted group at different position was also studied. Indoles with methyl substituent at C5 position (**2k** and **2l**) were obtained in slightly lower yields than at the C7 position (**2s** and **2t**), due to the steric of the 1-position of the indoline. Besides, a gram-scale reaction of **1q** (2.9 g, 10 mmol) was performed affording the product **2q** in an identical yield with the small-scale reaction (71% vs. 77%). In addition, the structure of **2r** was determined by X-ray crystallography [40].

To further examine the propensity for the reaction, indoles with different N-substituents were investigated (Table 4). Under the standard reaction conditions, the reaction proceeded smoothly and indoles bearing N-acetyl (**2v**) and N-Ts (**2w**) were also obtained in 81% and 75% yields, respectively.

To evaluate the synthetic utility of this novel method, we used it as the key step to build up the scaffold of the nonsteroidal anti-inflammatory drug molecule indomethacin (Scheme 1). When substrate **4** was subjected to the standard reaction conditions, the desired product **5** was obtained in 71% yield, which is the key intermediate of indomethacin. The following two steps to the final product indomethacin are described in a previous report [36]. Although indomethacin derivatives can be synthesized using Fisher indole synthesis and other cyclization methods, this methodology offers an alternative way to synthesize the key intermediate substituted N-benzoylindole **5** in one step, which is crucial to further diversity-oriented synthesis of analogs of indomethacin derivatives.

Table 3. Substrate scope of substrates with substituents at the positions (R^1) [a,b] Conditions: See Supplementary Materials for details.

[a] All reactions were carried out in 0.2 mmol scale. [b] Isolated yields. [c] Yield of the scale-up reaction (2.9 g, 10 mmol).

Table 4. Indoles with different N-substituents. [a,b] Conditions: See Supplementary Materials for details.

[a] All reactions were carried out in 0.2 mmol scale. [b] Yields referred to here are isolated yields.

On the basis of the previous studies, a plausible reaction mechanism is proposed (Scheme 2). Initially, the Pd^{II} catalyst first coordinates to the olefin to generate an intermediate **a**, followed by insertion of the alkene into the Pd^{II}-N bond in an amidopalladation reaction to give an intermediate **b**. Subsequent β-hydride elimination from the resulting alkyl-Pd^{II} species affords the intermediate c, which undergoes spontaneous isomerization–aromatization to form product **2a** [12,31]. Pd(0), in equilibrium with LPdH, was then reoxidized by the action of BQ.

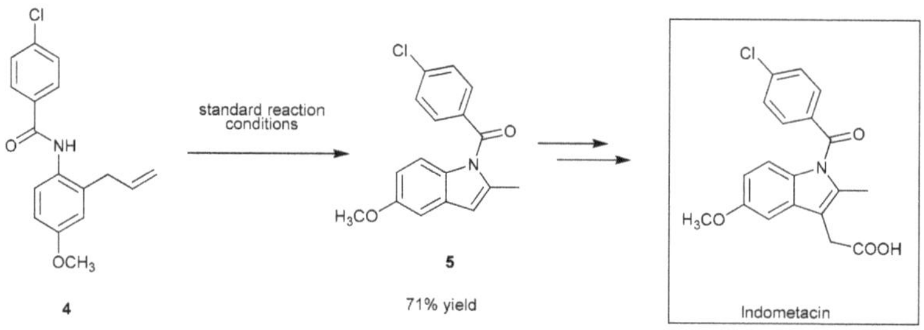

Scheme 1. Synthesis of indomethacin.

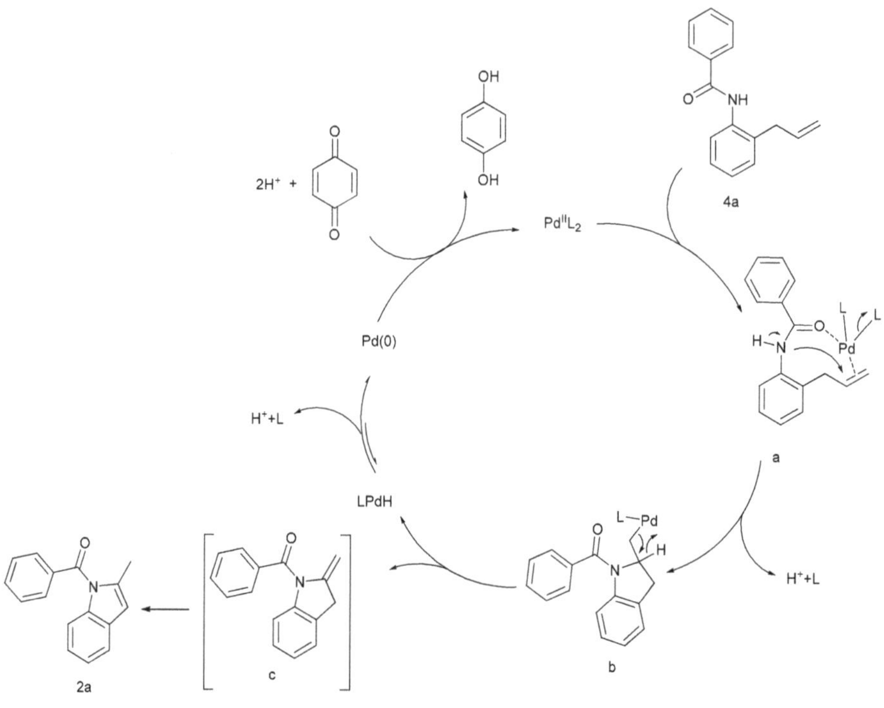

Scheme 2. Proposed Mechanism.

3. Materials and Methods

Column chromatography was performed on silica gel (Silica-P flash silica gel from Silicycle, size 40–63 µm). TLC was performed on silica gel 60/Kieselguhr F254. Mass spectra were recorded on an AEI-MS-902 mass spectrometer (EI+) or a LTQ Orbitrap XL (ESI+). ^{1}H, ^{13}C, and ^{19}F NMR were recorded on a Varian AMX400 (400, 100.6, and 376 MHz, respectively) or a Varian Unity Plus Varian-500 (500, 125, and 471 MHz, respectively). Chemical shift values for ^{1}H and ^{13}C NMR are reported in ppm with the solvent resonance as the internal standard (CHCl$_3$: δ 7.26 ppm for ^{1}H, δ77.0 ppm for ^{13}C). Data are reported as follows: chemical shifts, multiplicity (s = singlet, d = doublet, t = triplet, q = quartet, br = broad, m = multiplet), coupling constants (Hz), and integration. Melting points were determined on a Buchi B–545 melting point apparatus. All reactions were performed under anhydrous conditions and under N_2 atmosphere. All chemicals used were of analytical grade and were used as received without any further purification. All anhydrous solvents used in reactions were purchased in SureSeal bottles or dried over molecular sieves. Flash column chromatography was performed on Biotage Isolelera One with prepacked columns.

4. Conclusions

In summary, we developed an effective method for the synthesis of substituted *N*-benzoylindoles. Pd(II)-catalyzed synthesis of substituted *N*-benzoylindole was realized for the first time via C–H activation, starting from readily available substituent *N*-(2-allylphenyl) benzamide. Using inexpensive BQ as the oxidant, a series of substituted indoles were prepared in good to excellent yields under mild reaction conditions, which overcome the formation of byproduct benzoxazine. It should be noted that dibutyl phosphate (DBP) is the key to obtaining high yield and chemoselectivity in the present reaction. The indoles can be readily converted to many useful skeletons. As an example, this method was successfully used for the synthesis of a key skeleton of indomethacin.

Supplementary Materials: The following are available online, ^{1}H, ^{13}C, and ^{19}F-NMR spectra of compounds **1a–1w, 2a–2w, 4** and **5**.

Author Contributions: D.Z. and Z.C. conceived and designed the experiments. Z.C. and T.M. performed the experiments and mechanism studies. Y.Z. and Z.D.co-wrote the manuscript. H.Z. contributed to revising manuscript. All authors have read and agreed to the published version of the manuscript.

Funding: We are grateful for the support of this work by the program for Guangdong Introducing Innovative and Entrepreneurial Teams (2017ZT07C069), the National Natural Science Foundation of China (Nos. 21801260 and 21971267), and the "1000-Youth Talents Plan".

Conflicts of Interest: The authors declare no conflict of interest.

References and Notes

1. Somei, M.; Yamada, F. Simple indole alkaloids and those with a nonrearranged monoterpenoid unit. *Nat. Prod. Rep.* **2004**, *21*, 278–311. [CrossRef] [PubMed]
2. Kawasaki, T.; Higuchi, K. Simple indole alkaloids and those with a nonrearranged monoterpenoid unit. *Nat. Prod. Rep.* **2005**, *22*, 761–793. [CrossRef] [PubMed]
3. Kochanowska-Karamyan, A.J.; Hamann, M.T. Marine indole alkaloids: Potential new drug leads for the control of depression and anxiety. *Chem. Rev.* **2010**, *110*, 4489–4497. [CrossRef] [PubMed]
4. Randriambola, L.; Quirion, J.C.; Kan-Fan, C.; Husson, H.P. Structure of goniomitine, a new type of indole alkaloid. *Tetrahedron Lett.* **1987**, *28*, 2123–2126. [CrossRef]
5. Kaushik, N.K.; Kaushik, N.; Attri, P.; Kumar, N.; Kim, C.H.; Verma, A.K.; Choi, E.H. Biomedical importance of indoles. *Molecules* **2013**, *18*, 6620–6662. [CrossRef]
6. Fischer, E.; Jourdan, F. Ueber die Hydrazine der Brenztraubensä ure. *Ber. Dtsch. Chem. Ges.* **1883**, *16*, 2241–2245. [CrossRef]
7. Larock, R.C.; Yum, E.K. Synthesis of Indoles Via Palladium-Catalyzed Heteroannulation of Internal Alkynes. *J. Am. Chem. Soc.* **1991**, *113*, 6689–6690. [CrossRef]
8. Larock, R.C.; Yum, E.K.; Refvik, M.D. Synthesis of 2,3-disubstituted indoles via palladium-catalyzed annulation of internal alkynes. *J. Org. Chem.* **1998**, *63*, 7652–7662. [CrossRef]
9. Wagaw, S.; Yang, B.H.; Buchwald, S.L. A palladium-catalyzed strategy for the preparation of indoles: A novel entry into the Fischer indole synthesis. *J. Am. Chem. Soc.* **1998**, *120*, 6621–6622. [CrossRef]
10. Wagaw, S.; Yang, B.H.; Buchwald, S.L. A palladium-catalyzed method for the preparation of indoles via the Fischer indole synthesis. *J. Am. Chem. Soc.* **1999**, *121*, 10251–10263. [CrossRef]
11. Rutherford, J.L.; Rainka, M.P.; Buchwald, S.L. An annulative approach to highly substituted indoles: Unusual effect of phenolic additives on the success of the arylation of ketone enolates. *J. Am. Chem. Soc.* **2002**, *124*, 15168–15169. [CrossRef] [PubMed]

12. Hegedus, L.S.; Allen, G.F.; Waterman, E.L. Palladium assisted intramolecular amination of olefins. A new synthesis of indoles. *J. Am. Chem. Soc.* **1976**, *98*, 2674–2676. [CrossRef]
13. Hegedus, L.S.; Allen, G.F.; Bozell, J.J.; Waterman, E.L. Palladium-assisted intramolecular amination of olefins. Synthesis of nitrogen heterocycles. *J. Am. Chem. Soc.* **1978**, *100*, 5800–5807. [CrossRef]
14. Bozell, J.J.; Hegedus, L.S. Palladium-Assisted Functionalization of Olefins-a New Amination of Electron-Deficient Olefins. *J. Org. Chem.* **1981**, *46*, 2561–2563. [CrossRef]
15. Louis, S.H. Transition Metals in the Synthesis and Functionaiization of Indoles. *Angew. Chem. Int. Ed.* **1988**, *27*, 1113–1126. [CrossRef]
16. Stuart, D.R.; Bertrand-Laperle, M.; Burgess, K.M.; Fagnou, K. Indole synthesis via rhodium catalyzed oxidative coupling of acetanilides and internal alkynes. *J. Am. Chem. Soc.* **2008**, *130*, 16474–16475. [CrossRef]
17. Bernini, R.; Fabrizi, G.; Sferrazza, A.; Cacchi, S. Copper-catalyzed C-C bond formation through C-H functionalization: Synthesis of multisubstituted indoles from N-aryl enaminones. *Angew. Chem. Int. Ed.* **2009**, *48*, 8078–8081. [CrossRef]
18. Guan, Z.H.; Yan, Z.Y.; Ren, Z.H.; Liu, X.Y.; Liang, Y.M. Preparation of indoles via iron catalyzed direct oxidative coupling. *Chem. Commun.* **2010**, *46*, 2823–2825. [CrossRef]
19. Stuart, D.R.; Alsabeh, P.; Kuhn, M.; Fagnou, K. Rhodium(III)-catalyzed arene and alkene C-H bond functionalization leading to indoles and pyrroles. *J. Am. Chem. Soc.* **2010**, *132*, 18326–18339. [CrossRef]
20. Tan, Y.; Hartwig, J.F. Palladium-catalyzed amination of aromatic C-H bonds with oxime esters. *J. Am. Chem. Soc.* **2010**, *132*, 3676–3677. [CrossRef]
21. Huestis, M.P.; Chan, L.; Stuart, D.R.; Fagnou, K. The vinyl moiety as a handle for regiocontrol in the preparation of unsymmetrical 2,3-aliphatic-substituted indoles and pyrroles. *Angew. Chem. Int. Ed.* **2011**, *50*, 1338–1341. [CrossRef]
22. Neumann, J.J.; Rakshit, S.; Droge, T.; Wurtz, S.; Glorius, F. Exploring the oxidative cyclization of substituted N-aryl enamines: Pd-catalyzed formation of indoles from anilines. *Chem. Eur. J.* **2011**, *17*, 7298–7303. [CrossRef]
23. Wang, Y.; Ye, L.; Zhang, L. Au-catalyzed synthesis of 2-alkylindoles from N-arylhydroxylamines and terminal alkynes. *Chem. Commun.* **2011**, *47*, 7815–7817. [CrossRef]
24. Wei, Y.; Deb, I.; Yoshikai, N. Palladium-catalyzed aerobic oxidative cyclization of N-aryl imines: Indole synthesis from anilines and ketones. *J. Am. Chem. Soc.* **2012**, *134*, 9098–9101. [CrossRef]
25. Song, W.; Ackermann, L. Nickel-catalyzed alkyne annulation by anilines: Versatile indole synthesis by C-H/N-H functionalization. *Chem. Commun.* **2013**, *49*, 6638–6640. [CrossRef]
26. Wang, C.; Sun, H.; Fang, Y.; Huang, Y. General and efficient synthesis of indoles through triazene-directed C-H annulation. *Angew. Chem. Int. Ed.* **2013**, *52*, 5795–5798. [CrossRef]
27. Tremont, S.J.; Rahman, H.U. Ortho-Alkylation of Acetanilides Using Alkyl-Halides and Palladium Acetate. *J. Am. Chem. Soc.* **1984**, *106*, 5759–5760. [CrossRef]
28. Fix, S.R.; Brice, J.L.; Stahl, S.S. Efficient intramolecular oxidative amination of olefins through direct dioxygen-coupled palladium catalysis. *Angew. Chem. Int. Ed.* **2002**, *41*, 164–166. [CrossRef]
29. Nallagonda, R.; Rehan, M.; Ghorai, P. Synthesis of functionalized indoles via palladium-catalyzed aerobic oxidative cycloisomerization of o-allylanilines. *Org. Lett.* **2014**, *16*, 4786–4789. [CrossRef]
30. Ning, X.S.; Wang, M.M.; Qu, J.P.; Kang, Y.B. Synthesis of Functionalized Indoles via Palladium-Catalyzed Aerobic Cycloisomerization of o-Allylanilines Using Organic Redox Cocatalyst. *J. Org. Chem.* **2018**, *83*, 13523–13529. [CrossRef]
31. Savvidou, A.; IoannisTzaras, D.; Koutoulogenis, G.S.; Theodorou, A.; Kokotos, C.G. Synthesis of Benzofuran and Indole Derivatives Catalyzed by Palladium on Carbon. *Eur. J. Org. Chem.* **2019**, *2019*, 3890–3897. [CrossRef]
32. Cajaraville, A.; Lopez, S.; Varela, J.A.; Saa, C. Rh(III)-catalyzed tandem C-H allylation and oxidative cyclization of anilides: A new entry to indoles. *Org. Lett.* **2013**, *15*, 4576–4579. [CrossRef] [PubMed]
33. Kim, M.; Park, J.; Sharma, S.; Han, S.; Han, S.H.; Kwak, J.H.; Jung, Y.H.; Kim, I.S. Synthesis and C2-functionalization of indoles with allylic acetates under rhodium catalysis. *Org. Biomol. Chem.* **2013**, *11*, 7427–7434. [CrossRef] [PubMed]
34. Gong, T.-J.; Cheng, W.-M.; Su, W.; Xiao, B.; Fu, Y. Synthesis of indoles through Rh(III)-catalyzed C–H cross-coupling with allyl carbonates. *Tetrahedron Lett.* **2014**, *55*, 1859–1862. [CrossRef]

35. Andries-Ulmer, A.; Brunner, C.; Rehbein, J.; Gulder, T. Fluorine as a Traceless Directing Group for the Regiodivergent Synthesis of Indoles and Tryptophans. *J. Am. Chem. Soc.* **2018**, *140*, 13034–13041. [CrossRef]
36. Xia, H.D.; Zhang, Y.D.; Wang, Y.H.; Zhang, C. Water-Soluble Hypervalent Iodine(III) Having an I-N Bond. A Reagent for the Synthesis of Indoles. *Org. Lett.* **2018**, *20*, 4052–4056. [CrossRef]
37. McDonald, R.I.; Stahl, S.S. Modular synthesis of 1,2-diamine derivatives by palladium-catalyzed aerobic oxidative cyclization of allylic sulfamides. *Angew. Chem. Int. Ed.* **2010**, *49*, 5529–5532. [CrossRef]
38. Strambeanu, I.I.; White, M.C. Catalyst-controlled C-O versus C-N allylic functionalization of terminal olefins. *J. Am. Chem. Soc.* **2013**, *135*, 12032–12037. [CrossRef]
39. Osberger, T.J.; White, M.C. N-Boc amines to oxazolidinones via Pd(II)/bis-sulfoxide/Bronsted acid co-catalyzed allylic C-H oxidation. *J. Am. Chem. Soc.* **2014**, *136*, 11176–11181. [CrossRef]
40. CCDC 1894284 contains the supplementary crystallographic data for **2r**. These data can be obtained free of charge from The Cambridge Crystallographic Data Centre via www.ccdc.czm.ac.uk/datarequest/cif.

Sample Availability: Samples of the compounds are not available from the authors.

 © 2020 by the authors. Licensee MDPI, Basel, Switzerland. This article is an open access article distributed under the terms and conditions of the Creative Commons Attribution (CC BY) license (http://creativecommons.org/licenses/by/4.0/).

Article

Sulfoximines-Assisted Rh(III)-Catalyzed C–H Activation and Intramolecular Annulation for the Synthesis of Fused Isochromeno-1,2-Benzothiazines Scaffolds under Room Temperature

Bao Wang [1,2,3], Xu Han [1,2], Jian Li [1], Chunpu Li [1,2,*] and Hong Liu [1,2,3,*]

1. State Key Laboratory of Drug Research and CAS Key Laboratory of Receptor Research, Shanghai Institute of Materia Medica, Chinese Academy of Sciences, 555 Zu Chong Zhi Road, Shanghai 201203, China; wangbao@shanghaitech.edu.cn (B.W.); 201718012342004@simm.ac.cn (X.H.); jianl@simm.ac.cn (J.L.)
2. University of Chinese Academy of Sciences, No.19A Yuquan Road, Beijing 100049, China
3. School of Life Science and Technology, ShanghaiTech University, 100 Haike Road, Shanghai 201210, China
* Correspondence: lichunpu@simm.ac.cn (C.L.); hliu@simm.ac.cn (H.L.); Tel.: +86-50807042 (H.L.)

Academic Editor: Gianfranco Favi
Received: 14 May 2020; Accepted: 25 May 2020; Published: 28 May 2020

Abstract: A mild and facile Cp*Rh(III)-catalyzed C–H activation and intramolecular cascade annulation protocol has been proposed for the furnishing of highly fused isochromeno-1,2-benzothiazines scaffolds using S-phenylsulfoximides and 4-diazoisochroman-3-imine as substrates under room temperature. This method features diverse substituents and functional groups tolerance and relatively mild reaction conditions with moderate to excellent yields. Additionally, retentive configuration of sulfoximides in the conversion has been verified.

Keywords: sulfoximide; C–H activation; benzothiazine; rhodium

1. Introduction

Over the past decade, sulfoximines moiety has gained an increasing attention in organic chemistry [1–9] and pharmaceutical industries for their interesting properties such as multiple hydrogen-bond acceptor/donor functionalities, structural diversity, and favorable physicochemical properties [10–13]. For instances, sulfoximines along with benzothiazines scaffold-containing compounds possess diversified biologically active molecules such as antihypertensive activity α-adrenergic receptor blocker [14], anti-HIV nonnucleoside reverse transcriptase inhibitors [15], and hepatocytes protective mitogen-activated protein kinase kinase 4 (MKK4) inhibitor (Figure 1).

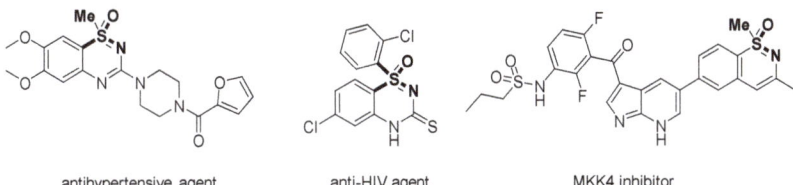

Figure 1. Representatives of biologically active compounds containing sulfoximines moiety.

Considering the significance of the sulfoximines motif as a pharmacophore in medicinal chemistry, synthetic methods accessing to this moiety have been increasingly studied. The typical route approach to the sulfoximines starting from commercially available sulfides requires two steps include oxidation and successively imination [6,16,17]. Sometimes, in order to produce the *N*-free sulfoximines, an additional step should be involved to dissociate the protecting group. Thus, these traditional methods have drawbacks such as requiring relatively harsh reaction conditions and poor step-economy for constructing cyclic sulfoximines. In recent years, transition-metal-catalyzed direct C–H functionalization has been broadly investigated, and these strategies have been proven to be an efficient tool for the rapid construction of C–C, C–O, C–N, or C–S bonds [18–30]. Additionally, owing to the high efficiency and easy accessibility of Rh(III) catalysis, the construction of 1,2-benzothiazines through Rh(III)-catalyzed C–H bond activation has attracted attention and been extensively studied [31–33]. For instance, in 2013, Bolm and coworkers developed a Rh(III)-catalyzed C–H functionalization for the synthesis of 1,2-benzothiazines starting from *N-H*-sulfoximines and alkynes (Scheme 1a) [34]. This process had been accomplished under 1 atm O_2 at 100 °C, and the desired products could be yielded in good to excellent yields but with limited structural diversity. Later in 2015, Bolm et al. has disclosed another strategy approaching the 1,2-benzothiazines via sulfoximines-assisted Rh(III)-catalyzed C–H functionalization and coupling reaction with diazo compounds (Scheme 1b) [35]. It is noteworthy that the dizao coupling partners should be electro-withdrawing groups-incorporated moieties, which lead to a limited versatility, and the process was carried out under argon at 100 °C. Recently, in Lee's group, an Rh(III)-catalyzed domino C–H activation/cyclization strategy has been reported by mixing sulfoximine and pyridotriazole compounds to build the 1,2-benzothiazines skeleton (Scheme 1c) [36]. In the methodology, the target products were 1,2-benzothiazines bearing pyridyl motifs as well as carbonyl groups and the process was conducted at a high reaction temperature. However, it is of note that most of these strategies have to be conducted at harsh reaction conditions with a limited reaction versatility and structural diversity. Only in 2019, Wu et al. reported a mild protocol to synthesize 1,2-benzothiazines derivatives [37]. Moreover, the constructed 1,2-benzothiazines scaffolds were bicyclic moieties, and to the best of our knowledge, only limited examples have been disclosed for the synthesis of fused 1,2-benzothiazines motif under harsh reaction conditions [38].

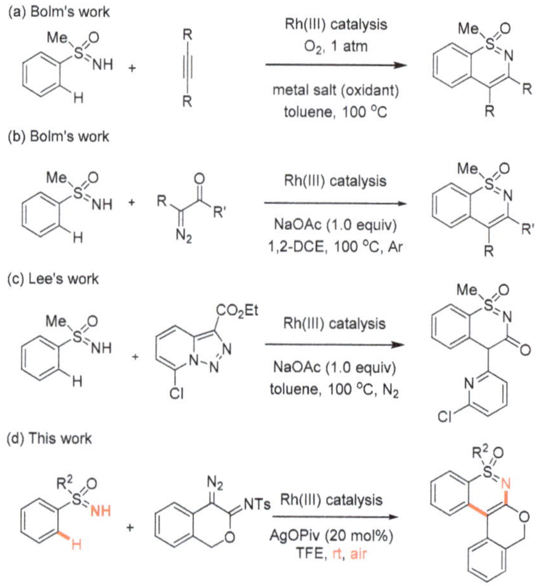

Scheme 1. Synthetic methods access to sulfoximines. (**a**) Bolm's work; (**b**) Bolm's work; (**c**) Lee's work; (**d**) This work.

In continuation of our studies on the establishment of fused ring heterocyclic compounds via Rh(III)-catalyzed C-H functionalization [39–41], we envisioned that a fused 1,2-benzothiazines scaffold could be achieved via an Rh(III)-catalyzed C–H functionalization/annulation between sulfoximines and 4-diazoisochroman-3-imines. To our delight, the fused isochromeno-1,2-benzothiazines were successfully accomplished in good to excellent yields through a redox-neutral process, and our strategy could be carried out in the air under room temperature with broad generality and versatility. We herein described our results in detail.

2. Results and Discussion

Based on our previous work and relevant reports, we initially focused the studies on the Rh(III)-catalyzed coupling of sulfoximide (**1a**) and 4-diazoisochroman-3-imine (**2a**) (Table 1). In the presence of 10 mol% [Cp*RhCl$_2$]$_2$ and 40 mol% AgSbF$_6$ in dichloroethane (DCE) at 80 °C, the desired product **3aa** was produced in 16% yield, and its structure was further confirmed by ^{1}H NMR spectroscopy (Table 1, Entry 2). We further assessed the reaction temperature in different solvents (Table 1, Entries 1–5), and the results demonstrated that when applying hexafluoro-isopropanol (HFIP) as the solvent, the reaction could be preceded by producing the **3aa** in 47% yield under room temperature. This result leads us to realize that the polar alcohols would more facilitate the reaction at mild conditions rather than alkane such as DCE, which might retarded the process (Table 1, Entries 4 and 5). Inspired by this result, the other reaction solvents were screened under room temperature (Table 1, Entries 6–9). After a series of solvents was examined, the trifluoroethanol (TFE) was found to be the optimal in which **3aa** was obtained in 75% yield (Table 1, Entry 6), and the structure of the desired product was further verified by X-ray crystallography (CCDC 1988905). We next investigated the effect of additives and gratifyingly, AgOPiv emerged to be the most effective additive among all those examined ones (Table 1, entries 10–13) to afford the desired product in an excellent isolated yield of 92%. Afterward, the ratio of the catalyst and additive was subsequently conducted (Table 1, Entries 14–16). The results disclosed that the 1:4 ratio of catalyst and additive was necessary for the full conversion of the starting material **1a** and **2a** (Table 1, Entries 13–17). In addition, further exploration proved that there was no obvious discrepancy of the yield of **3aa** when the reaction was moved to an argon atmosphere (Table 1, Entry 18). However, when the reaction time was shortened to 12 h, the yield of **3aa** was slightly reduced (Table 1, Entry 19). Moreover, we also tested different transition-metal catalysts such as the cost-effective ruthenium (II) or iridium (III) complex, and **3aa** could be only detected in 85% yield when treated with 10 mol% [Cp*IrCl$_2$]$_2$ as catalyst (Table 1, entries 20–21). Notably, **3aa** could not be detected when in the absence of [Cp*RhCl$_2$]$_2$, which indicated the [Cp*RhCl$_2$]$_2$ catalyst is indispensable for this transformation (Table 1, Entry 22). Therefore, the standard conditions for this Rh(III)-catalyzed coupling reaction is 5 mol %[Cp*RhCl$_2$]$_2$ and 20 mol %AgOPiv in TFE at room temperature for 18 h in the air.

After establishing the optimal reaction conditions, we next turned to investigate the scope of sulfoximide derivatives. As illustrated in Table 2, this Rh-catalyzed coupling reaction could proceed smoothly with substituted S-aryl sulfoximine substrates bearing electron-donating or electron-withdrawing substituents, and the corresponding products could be yielded in moderate to good yields. Sulfoximines with substituents including methyl, methoxyl, halogen, nitro, etc. installed at the *para*-position of the benzene ring coupled with **2a** smoothly to afford the isochromeno-benzothiazines products in good yields (**3ca–3ha**), albeit methyl substituted product (**3ba**) was obtained in only 47% yield. It should be mentioned that functional groups such as carbonyl, ester, or carbamate were all viable under the standard reaction conditions to give the desired products in 66% (**3ia**), 82% (**3ja**), and 89% (**3ka**) yields, respectively. This result indicated that the strategy could be broadly extended for further transformation. The *ortho*-methyl-incorporated sulfoximines also gave a relatively low yield of the desired product (**3la**, 58% yield). Notably, the substituents of sulfoximine substrates at different positions of its benzene ring did not alter the reaction efficiency, as for the halogenated substrates (compare **3ga**, **3ma**, and **3na**) provided the desired products in similar yields (72% to 74%). Interestingly, when an electron-withdrawing group such as bromo was installed at the *meta*-position of the benene ring in **1a**, only the *o*-C-H bond of sulfoximine located on the less hindered site was activated, which led to a production of **3na** as a single isomer in 72% yield. On the contrary, if an electron-donating group was incorporated on the meta-position, the products were isolated as a regioisomer (**3oa** and **3oa′**) in a total yield of 50%. In addition, naphthalene-fused sulfoximide was also coupled with **2a** smoothly, and the desired product **3pa** was obtained as a single isomer through activation/annulation on the less hindered site of sulfoximine in 73% yield.

Table 1. Optimization of reaction conditions [a]. TFE: trifluoroethanol.

Entry	Catalyst (mol%)	Additive (mol%)	Solvent	Temp (°C)	Yield of 3aa (%) [b]
1	[Cp*RhCl$_2$]$_2$ (10)	AgSbF$_6$ (40)	DCE	100	ND
2	[Cp*RhCl$_2$]$_2$ (10)	AgSbF$_6$ (40)	DCE	80	16
3	[Cp*RhCl$_2$]$_2$ (10)	AgSbF$_6$ (40)	HFIP	80	trace
4	[Cp*RhCl$_2$]$_2$ (10)	AgSbF$_6$ (40)	DCE	rt	trace
5	[Cp*RhCl$_2$]$_2$ (10)	AgSbF$_6$ (40)	HFIP	rt	47
6	[Cp*RhCl$_2$]$_2$ (10)	AgSbF$_6$ (40)	TFE	rt	75
7	[Cp*RhCl$_2$]$_2$ (10)	AgSbF$_6$ (40)	EtOH	rt	57
8	[Cp*RhCl$_2$]$_2$ (10)	AgSbF$_6$ (40)	THF	rt	ND
9	[Cp*RhCl$_2$]$_2$ (10)	AgSbF$_6$ (40)	DME	rt	trace
10	[Cp*RhCl$_2$]$_2$ (10)	AgOAc (40)	TFE	rt	69
11	[Cp*RhCl$_2$]$_2$ (10)	AgNTf$_2$ (40)	TFE	rt	27
12	[Cp*RhCl$_2$]$_2$ (10)	AgBF$_4$ (40)	TFE	rt	ND
13	[Cp*RhCl$_2$]$_2$ (10)	AgOPiv (40)	TFE	rt	99 (92) [c]
14	[Cp*RhCl$_2$]$_2$ (20)	AgOPiv (40)	TFE	rt	47
15	[Cp*RhCl$_2$]$_2$ (5)	AgOPiv (40)	TFE	rt	77
16	[Cp*RhCl$_2$]$_2$ (5)	AgOPiv (20)	TFE	rt	97 (93) [c]
17	[Cp*RhCl$_2$]$_2$ (2.5)	AgOPiv (10)	TFE	rt	87
18 [d]	[Cp*RhCl$_2$]$_2$ (10)	AgOPiv (40)	TFE	rt	90 (89) [c]
19 [e]	[Cp*RhCl$_2$]$_2$ (10)	AgOPiv (40)	TFE	rt	82
20	[Cp*RuCl$_2$]$_2$ (10)	AgOPiv (40)	TFE	rt	trace
21	[Cp*IrCl$_2$]$_2$ (10)	AgOPiv (40)	TFE	rt	85
22	-	AgOPiv (40)	TFE	rt	ND

[a] Reaction conditions: **1a** (0.15 mmol), **2a** (0.165 mmol), catalyst and additive in solvent (2.5 mL) under air. [b] Determined by ^{1}H NMR spectroscopy using 1,3,5-trimethoxybenzene as an internal standard. [c] Isolated yield in parentheses. [d] Under an Argon atmosphere. [e] The reaction time was shortened to 12 h.

We also evaluated a variety of *S*-substituted phenylsulfoximine substrates (**3qa**–**3va**), and the results revealed that alkyl, halogen, hydroxyl, and aryl substituents on sulfur substrates were also compatible under the standard conditions and generated the corresponding product in moderate to good yields varying from 40% to 90%. In particularly, the cyclopropyl with huge steric hindrance effect exhibited no impact on the reaction efficiency and showed the best reactivity with 90% yield. Additionally, pyridine sulfoximine substrate could not generate the desired product (**3wa**), which might be caused by the strong coordination effect of the nitrogen atom in pyridine, which ceased the reaction process.

Table 2. Substrate scope of sulfoximines [a].

Product	Yield
R = H, **3aa**	93%
R = Me, **3ba**	47%
R = OMe, **3ca**	76%
R = CH$_2$OH, **3da**	78%
R = F, **3ea**	80%
R = Cl, **3fa**	80%
R = Br, **3ga**	74%
R = NO$_2$, **3ha**	77%
R = Ac, **3ia**	66%
R = CH$_2$COOEt, **3ja**	82%
R = NHBoc, **3ka**	89%
3la	58%
3ma	73%
3na	72%
3oa + 3oa' (3oa:3oa' = 2.7:1)[b]	50%
3pa	73%
3qa	56%
3ra	90%
3sa	77%
3ta	42%
3ua	79%
3va	48%
3wa	NR[c]

[a] Reaction conditions: **1** (0.15 mmol), **2a** (0.165 mmol), [Cp*RhCl$_2$]$_2$ (5 mol%) and AgOPiv (20 mol%) in TFE (2.5 mL) under air at room temperature for 18 h. All listed yields are isolated ones. [b] Determined by ^{1}H NMR spectroscopy. [c] NR means No Reaction.

Subsequently, we investigated the scope of 4-diazoisochroman-3-imine coupling partner (Table 3). The introduction of both electron-donating or electron-withdrawing substituents including methyl, methoxy, Cl, F, and trifluoromethyl at the 6- or 7- positions of isochroman were all tolerated in this coupling reaction (**3ab**–**3aj**) and the yields were varying from 53% to 76%. Among all the substituents, the electron-deficient trifluoromethyl isochroman substrates exhibited good reactivity and independently furnished the products **3aj** and **3ai** in 76% and 74% yields. Besides, the nitro group-substituted substrate was not compatible for this coupling reaction, and no desired product (**3ak**) was achieved. Substituents at different positions has no obvious influence on the reaction efficiency, as the yields of the 6- and 7- substituted substrates are basically the same.

Table 3. Substrate scope of 4-diazoisochroman-3-imines [a].

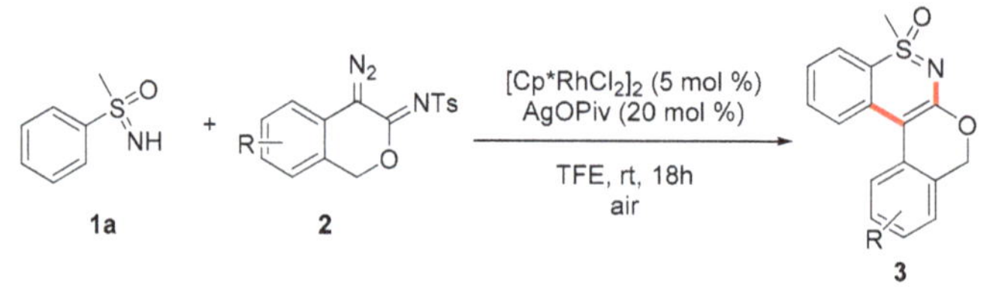

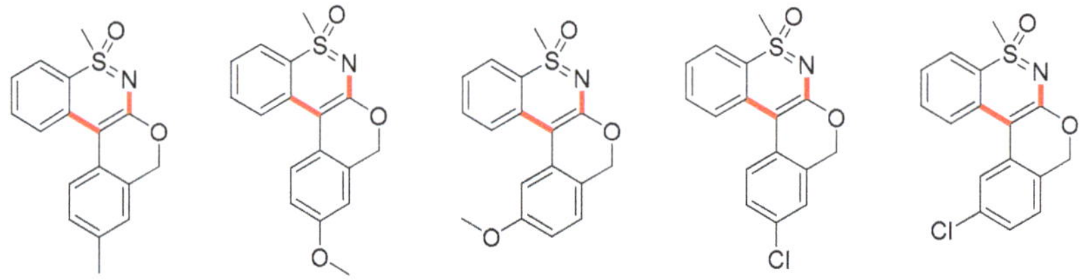

Compound	Yield
3ab	63%
3ac	53%
3ad	72%
3ae	69%
3af	57%
3ag	60%
3ah	68%
3ai	76%
3aj	74%
3ak	NR [b]

[a] Reaction conditions: **1a** (0.15 mmol), **2** (0.165 mmol), [Cp*RhCl$_2$]$_2$ (5 mol%) and AgOPiv (20 mol%) in TFE (2.5 mL) under air at room temperature for 18 h. All listed yields are isolated ones. [b] NR means No Reaction.

The chirality of the sufoximine group is important [42–46]. In order to verify the stereospecificity in the whole reaction process, optically pure *R*- and *S*-configured **1a** were parallelly coupled with **2a** under the standard condition (Scheme 2). The results demonstrated that the corresponding products were obtained with a retention of configuration with no erosion of the enantiopurity of the sulfoximine occurred, which indicated that the current coupling protocol possesses a potential utility on asymmetric synthesis.

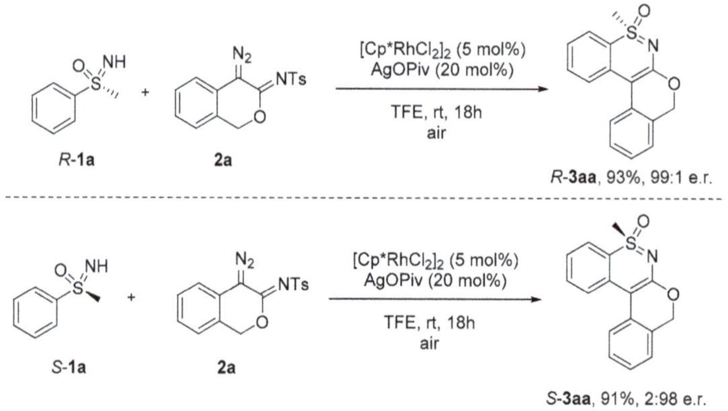

R-**1a** + **2a** → *R*-**3aa**, 93%, 99:1 e.r.

S-**1a** + **2a** → *S*-**3aa**, 91%, 2:98 e.r.

Scheme 2. Conversion of stereoisomer substrates.

Next, we conducted a series of control experiments to explore the preliminary reaction mechanism (Scheme 3). First, the kinetic isotope effect (KIE) experiment in intramolecular between d^1-**1a** and **2a** were performed and a small KIE value (k_H/k_D) was measured as 0.47, which indicated that the aryl Csp^2-H bond cleavage was not the rate-limiting step (Scheme 3a) [47]. Next, in the H/D exchange study, phenylsulfoximine **1a** was conducted under the standard conditions in deuterium TFE in the absence of **2a**, and the deuterated d^2-**1a** recovered in 91% deuterium at the *ortho*-position of benzene, which suggested that the C–H bond activation process was reversible (Scheme 3b). Finally, an intermolecular competitive experiment between electron-rich and electron-deficient substrates (**1c/1i** = 1:1) lead to the products **3ca/3ia** with a ratio of 1.46, implying that an electrophilic rhodation of C–H bond activation probably involved in the catalytic cycle (Scheme 3c).

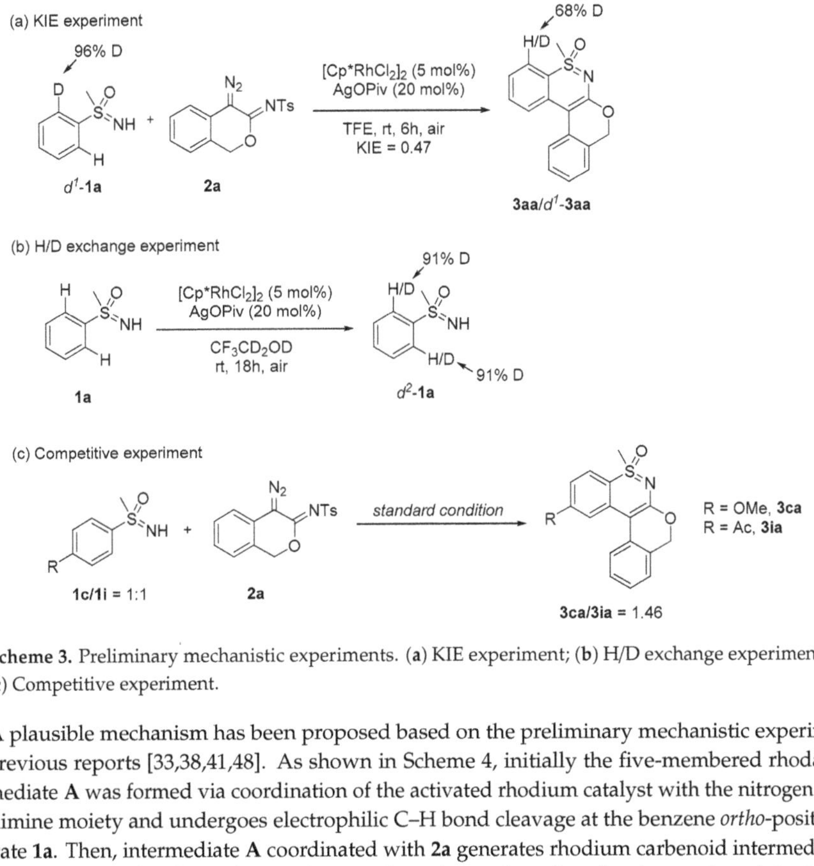

Scheme 3. Preliminary mechanistic experiments. (a) KIE experiment; (b) H/D exchange experiment; (c) Competitive experiment.

A plausible mechanism has been proposed based on the preliminary mechanistic experiments and previous reports [33,38,41,48]. As shown in Scheme 4, initially the five-membered rhodacycle intermediate **A** was formed via coordination of the activated rhodium catalyst with the nitrogen of the sulfoximine moiety and undergoes electrophilic C–H bond cleavage at the benzene *ortho*-position of substrate **1a**. Then, intermediate **A** coordinated with **2a** generates rhodium carbenoid intermediate **B**, followed by intramolecular carbene migratory insertion and produced a six-membered rhodacycle intermediate **C**. Finally, the intermediate **C** protonation to yield compound **D** and regenerated the active cationic Cp*Rh(III) species for the next catalytic cycle. Compound **D** could easily undergo intramolecular nucleophilic attack of imine and elimination of the *p*-toluenesulfonamide (TsNH$_2$) process to produce the desired isochromeno-1,2-benzothiazines product **3aa**.

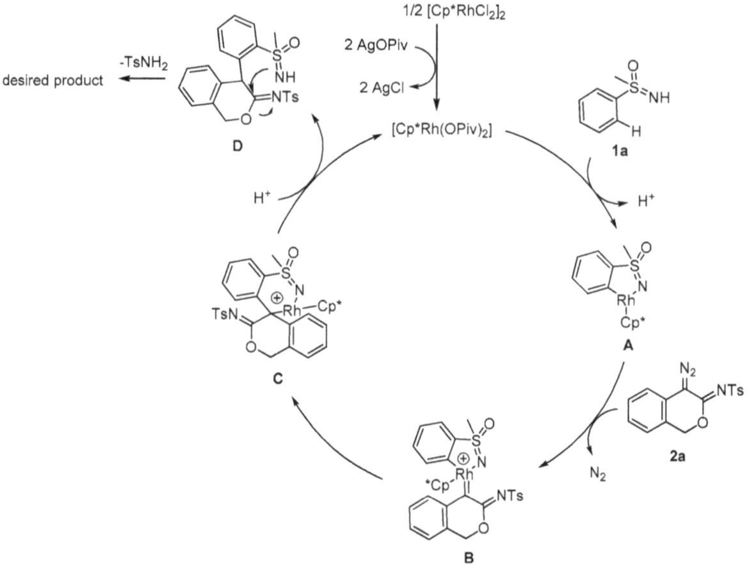

Scheme 4. Proposed mechanism.

3. Materials and Methods

3.1. General Information

All reagents and solvents were purchased from commercial sources (J&K Scientific Co., Ltd., Beijing, China; TCI Development Co., Ltd., Shanghai, China; Adamas Reagent, Co., Ltd., Shanghai, China.) and used without further purification. The analytical thin layer chromatography (TLC) was HSGF 254 (0.15–0.2 mm thickness). All products were characterized by their NMR and MS spectra. ^{1}H and ^{13}C nuclear magnetic resonance spectra (NMR) were acquired on a Bruker 400 MHz or 500 MHz or 600 MHz NMR spectrometer (Billerica, MA, USA). Chemical shifts were reported in parts per million (ppm, δ) downfield from tetramethylsilane, and the coupling constants (J) were indicated in Hz. Proton coupling patterns are described as singlet (s), doublet (d), triplet (t), quartet (q), multiplet (m), doublet of doublets (dd), and broad (br). Low-resolution mass spectra (LRMS) data were measured on Agilent 1260 Infinity II (Palo Alto, CA, USA) with Electrospray Ionization (ESI). High-resolution mass spectra (HRMS) data were measured on an Agilent G6520 Q-TOF (Palo Alto, CA, USA) with Electrospray Ionization (ESI). AgOPiv was prepared according to the reported literature method [49].

3.2. Experimental Part Method

3.2.1. General Procedure A for the Synthesis of Substrates **1a–1w**

To a stirred solution of sulfide (1 mmol) in MeOH (10 mL) was added (NH$_4$)$_2$CO$_3$ (1.5 equiv.). Subsequently, PhI(OAc)$_2$ (2.3 equiv.) was added, and the solution was stirred at rt for 10 min. The solvent was removed under reduced pressure, and the crude product was purified by flash column chromatography eluted with dichloromethane (DCM)/MeOH from 30:1 to 10:1 to give the desired product **1**. Compounds **1e**, **1f**, **1q**, *R*-**1a**, and *S*-**1a** were purchased from commercial sources and used without further purification. Compounds **1a–1c**, **1g-1i**, **1l–1p**, **1r–1w** are known compounds.

[4-(*S*-Methylsulfonimidoyl)phenyl]methanol (**1d**): white solid; m.p.: 108–109 °C; ^{1}H NMR (500 MHz, DMSO-d_6): δ 7.88 (d, *J* = 8.3 Hz, 2H), 7.55–7.49 (m, 2H), 5.41 (t, *J* = 5.7 Hz, 1H), 4.59 (d, *J* = 5.6 Hz, 2H),

4.15 (s, 1H), 3.04 (s, 3H); ^{13}C NMR (126 MHz, DMSO-d_6): δ 147.5, 142.3, 127.2, 126.6, 62.2, 46.0; LRMS (ESI): *m/z* 186.0 [M + H]$^+$; HRMS (ESI): calculated for $C_8H_{12}NO_2S$ [M + H]$^+$: 186.0583, found: 186.0584.

Ethyl [4-(S-methylsulfonimidoyl)phenyl]acetate (**1j**): colorless oil (207.5 mg, 86% yield); ^{1}H NMR (600 MHz, DMSO-d_6): δ 7.88 (d, *J* = 8.3 Hz, 2H), 7.49 (d, *J* = 8.4 Hz, 2H), 4.18 (s, 1H), 4.09 (q, *J* = 7.1 Hz, 2H), 3.80 (s, 2H), 3.05 (s, 3H), 1.19 (t, *J* = 7.1 Hz, 3H); ^{13}C NMR (126 MHz, CDCl$_3$): δ 170.6, 142.5, 139.8, 130.4, 128.1, 61.4, 46.3, 41.2, 14.3; LRMS (ESI): *m/z* 242.1 [M + H]$^+$; HRMS (ESI): calculated for $C_{11}H_{16}NO_3S$ [M + H]$^+$: 242.0845, found: 242.0841.

tert-Butyl (4-(S-methylsulfonimidoyl)phenyl)carbamate (**1k**): white solid; m.p.: 155–156 °C; ^{1}H NMR (600 MHz, DMSO-d_6): δ 9.78 (s, 1H), 7.80 (d, *J* = 8.7 Hz, 2H), 7.63 (d, *J* = 8.5 Hz, 2H), 4.01 (s, 1H), 3.00 (s, 3H), 1.49 (s, 9H); ^{13}C NMR (126 MHz, DMSO-d_6): δ 152.5, 143.3, 136.7, 128.4, 117.5, 79.8, 46.2, 28.0; LRMS (ESI) *m/z*: 271.1 [M + H]$^+$; HRMS (ESI) *m/z*: calculated for $C_{12}H_{18}N_2O_3S$ [M + H]$^+$: 271.1111, found: 271.1115.

3.2.2. General Procedure for the Synthesis of Substrates **2a–2k**

To a two-neck round bottom flask was successively added (2-ethynylphenyl)methanols (5.0 mmol, 1.0 equiv.), CuBr (0.5 mmol, 0.1 equiv.), and Et$_3$N (10.0 mmol, 2.0 equiv.) in anhydrous MeCN (50 mL, 0.1 M). The mixture was evacuated and refilled with Ar 3 times. To the resulting mixture was slowly added *p*-toluenesulfonyl azide (75% in EA solution, 11.0 mmol, 2.2 equiv.) over 10 min under Ar, and the reaction was processed under room temperature for 4–6 h. Then, the mixture was filtered through a pad of celite and washed with DCM (30 mL × 3). The combined organic layer was washed with saturated NaHCO$_3$, dried over anhydrous Na$_2$SO$_4$, filtered, and concentrated under reduced pressure. The crude residue was purified by flash chromatography eluted with hexane/EA/DCM = 5:1:2 to give the desired product **2** as a yellow solid. Compounds **2a–2k** are known compounds.

3.2.3. General Procedure for the Synthesis of Compounds **3**

To a 15 mL vial was added sulfoximines **1** (0.15 mmol), 4-diazoisochroman-3-imines **2** (0.165 mmol), [Cp*RhCl$_2$]$_2$ (5 mol%), and AgOPiv (20 mol%) under air. Trifluoroethanol (TFE, 2.5 mL) was added subsequently. The resulting mixture was stirred at ambient temperature for 18 h. Upon completion of the reaction, the mixture was filtered through a celite pad and washed with DCM (10 mL × 3). The combined organic layer was concentrated under vacuo, and the residue was purified by silica gel chromatography eluting with DCM/MeOH from 50:1 to 10:1 to give the desired isochromeno-1,2-benzothiazines product **3**.

3.2.4. Characterization of the Products

5-Methyl-8H-5λ^4-isochromeno[3,4-*c*][1,2]benzothiazine 5-oxide (**3aa**): yellow-green solid; m.p.: 182–184 °C; ^{1}H NMR (500 MHz, CDCl$_3$): δ 8.07 (d, *J* = 9.6 Hz, 1H), 7.78 (dd, *J* = 8.0, 1.3 Hz, 1H), 7.62–7.54 (m, 2H), 7.38–7.26 (m, 2H), 7.21–7.11 (m, 2H), 5.20–4.99 (m, 2H), 3.49 (s, 3H); ^{13}C NMR (126 MHz, CDCl$_3$): δ 157.3, 135.0, 132.9, 131.3, 128.8, 128.2, 125.0, 124.8, 124.5, 124.4, 123.2, 122.2, 120.0, 91.9, 70.3, 43.0; LRMS (ESI): *m/z* 284.1 [M + H]$^+$; HRMS (ESI): calculated for $C_{16}H_{14}NO_2S$ [M + H]$^+$: 284.0740, found: 284.0745.

2,5-Dimethyl-8H-5λ^4-isochromeno[3,4-*c*][1,2]benzothiazine 5-oxide (**3ba**): yellow solid; m.p.: 101–103 °C; ^{1}H NMR (600 MHz, CDCl$_3$): δ 7.87 (s, 1H), 7.67 (d, *J* = 8.2 Hz, 1H), 7.61 (d, *J* = 8.8 Hz, 1H), 7.32 (t, *J* = 6.8 Hz, 1H), 7.20 (d, *J* = 7.4 Hz, 1H), 7.18–7.13 (m, 2H), 5.14–5.00 (m, 2H), 3.46 (s, 3H), 2.44 (s, 3H); ^{13}C NMR (151 MHz, CDCl$_3$): δ 157.3, 143.7, 135.2, 131.4, 128.8, 128.2, 125.7, 125.0, 124.7, 124.2, 123.2, 122.2, 117.6, 91.6, 70.2, 43.1, 22.2; LRMS (ESI): *m/z* 298.0 [M + H]$^+$; HRMS (ESI): calculated for $C_{17}H_{16}NO_2S$ [M + H]$^+$: 298.0896, found: 298.0900.

2-Methoxy-5-methyl-8H-5λ^4-isochromeno[3,4-*c*][1,2]benzothiazine 5-oxide (**3ca**): pale yellow solid; m.p.: 105–106 °C; ^{1}H NMR (400 MHz, CDCl$_3$): δ 7.70 (d, *J* = 8.9 Hz, 1H), 7.64 (d, *J* = 7.8 Hz, 1H), 7.47

(d, *J* = 2.4 Hz, 1H), 7.35–7.27 (m, 1H), 7.20 (d, *J* = 5.9 Hz, 1H), 7.14 (t, *J* = 7.3 Hz, 1H), 6.88 (dd, *J* = 8.9, 2.4 Hz, 1H), 5.07 (q, *J* = 12.2 Hz, 2H), 3.86 (s, 3H), 3.42 (s, 3H); ^{13}C NMR (126 MHz, CDCl$_3$): δ 163.3, 157.9, 137.5, 131.5, 128.9, 128.2, 125.5, 125.1, 124.6, 121.8, 112.9, 112.8, 106.7, 91.5, 70.1, 55.7, 43.7; LRMS (ESI): *m/z* 314.0 [M + H]$^+$; HRMS (ESI): calculated for C$_{17}$H$_{16}$NO$_3$S [M + H]$^+$: 314.0845, found: 314.0844.

(5-Methyl-5-oxido-8H-5λ^4-isochromeno[3,4-*c*][1,2]benzothiazin-2-yl)methanol (**3da**): orange solid; m.p.: 65–67 °C; ^{1}H NMR (600 MHz, CDCl$_3$): δ 7.99 (s, 1H), 7.66 (d, *J* = 8.2 Hz, 1H), 7.54 (d, *J* = 7.8 Hz, 1H), 7.29–7.23 (m, 2H), 7.13 (d, *J* = 6.8 Hz, 2H), 5.01–4.93 (m, 2H), 4.73 (s, 2H), 3.40 (s, 3H), 2.97 (s, 1H); ^{13}C NMR (126 MHz, CDCl$_3$): δ 157.2, 146.6, 135.0, 131.1, 128.7, 128.3, 125.0, 124.8, 123.4, 122.8, 122.1, 121.7, 118.6, 92.0, 70.1, 64.5, 42.9; LRMS (ESI): *m/z* 314.0 [M + H]$^+$; HRMS (ESI): calculated for C$_{17}$H$_{16}$NO$_3$S [M + H]$^+$: 314.0845, found: 314.0846.

2-Fluoro-5-methyl-8H-5λ^4-isochromeno[3,4-*c*][1,2]benzothiazine 5-oxide (**3ea**): yellow solid; m.p.: 149–151 °C; ^{1}H NMR (600 MHz, CDCl$_3$): δ 7.79 (dd, *J* = 8.9, 5.5 Hz, 1H), 7.70 (dd, *J* = 11.3, 2.5 Hz, 1H), 7.57 (d, *J* = 7.6 Hz, 1H), 7.33 (td, *J* = 7.5, 1.7 Hz, 1H), 7.22–7.13 (m, 2H), 7.07–6.99 (m, 1H), 5.15–5.02 (m, 2H), 3.47 (s, 3H); ^{13}C NMR (151 MHz, CDCl$_3$): δ 157.8, 134.5, 133.5, 132.2, 131.9, 131.1–130.2 (m), 125.5, 125.0, 124.0, 123.4, 121.5–121.3 (m), 120.2, 118.6 (q, *J* = 3.8 Hz), 91.3, 69.7, 43.0; LRMS (ESI): *m/z* 302.0 [M + H]$^+$; HRMS (ESI): calculated for C$_{16}$H$_{13}$FNO$_2$S [M + H]$^+$: 302.0646, found: 302.0650.

2-Chloro-5-methyl-8H-5λ^4-isochromeno[3,4-*c*][1,2]benzothiazine 5-oxide (**3fa**): yellow-green solid; m.p.: 210–211 °C; ^{1}H NMR (600 MHz, CDCl$_3$): δ 8.04 (d, *J* = 2.0 Hz, 1H), 7.70 (d, *J* = 8.6 Hz, 1H), 7.57 (d, *J* = 7.8 Hz, 1H), 7.39–7.30 (m, 1H), 7.28 (d, *J* = 1.9 Hz, 1H), 7.22–7.15 (m, 2H), 5.15–5.02 (m, 2H), 3.48 (s, 3H); ^{13}C NMR (151 MHz, CDCl$_3$): δ 157.6, 139.2, 136.0, 130.2, 128.2, 127.9, 124.6, 124.6, 124.3, 124.0, 123.2, 121.5, 117.3, 91.1, 69.8, 42.6; LRMS (ESI): *m/z* 318.0 [M + H]$^+$; HRMS (ESI): calculated for C$_{16}$H$_{13}$ClNO$_2$S [M + H]$^+$: 318.0350, found: 318.0350.

2-Bromo-5-methyl-8H-5λ^4-isochromeno[3,4-*c*][1,2]benzothiazine 5-oxide (**3ga**): yellow-green solid; m.p.: 93–94 °C; ^{1}H NMR (500 MHz, CDCl$_3$): δ 8.21 (d, *J* = 1.8 Hz, 1H), 7.62 (d, *J* = 8.5 Hz, 1H), 7.56 (d, *J* = 7.8 Hz, 1H), 7.42 (dd, *J* = 8.5, 1.8 Hz, 1H), 7.37–7.33 (m, 1H), 7.22–7.15 (m, 2H), 5.16–5.02 (m, 2H), 3.48 (s, 3H).; ^{13}C NMR (126 MHz, CDCl$_3$): δ 157.6, 136.1, 130.1, 128.2, 128.0, 127.8, 126.8, 126.3, 124.6, 124.6, 124.2, 121.5, 117.7, 91.0, 69.8, 42.5; LRMS (ESI): *m/z* 361.0 [M + H]$^+$; HRMS (ESI): calculated for C$_{16}$H$_{13}$BrNO$_2$S [M + H]$^+$: 361.9845, found: 361.9850.

5-Methyl-2-nitro-8H-5λ^4-isochromeno[3,4-*c*][1,2]benzothiazine 5-oxide (**3ha**): brown solid; m.p.: 179–181 °C; ^{1}H NMR (600 MHz, CDCl$_3$): δ 8.92 (d, *J* = 2.2 Hz, 1H), 8.04 (dd, *J* = 8.7, 2.1 Hz, 1H), 7.91 (d, *J* = 8.7 Hz, 1H), 7.56 (d, *J* = 7.8 Hz, 1H), 7.39 (dt, *J* = 8.2, 4.4 Hz, 1H), 7.23 (d, *J* = 4.1 Hz, 2H), 5.21–5.03 (m, 2H), 3.61 (s, 3H); ^{13}C NMR (126 MHz, CDCl$_3$): δ 158.1, 150.0, 135.5, 129.6, 128.3, 128.1, 125.3, 124.8, 124.3, 121.8, 121.4, 119.5, 117.3, 92.6, 70.0, 42.2; LRMS (ESI): *m/z* 329.2 [M + H]$^+$; HRMS (ESI): calculated for C$_{16}$H$_{13}$N$_2$O$_4$S [M + H]$^+$: 329.0591, found: 329.0603.

1-(5-Methyl-5-oxido-8H-5λ^4-isochromeno[3,4-*c*][1,2]benzothiazin-2-yl)ethanone (**3ia**): orange-yellow solid; m.p.: 112–114 °C; ^{1}H NMR (600 MHz, CDCl$_3$): δ 8.63 (s, 1H), 7.86–7.80 (m, 2H), 7.57 (d, *J* = 7.8 Hz, 1H), 7.35 (td, *J* = 7.5, 1.7 Hz, 1H), 7.24–7.16 (m, 2H), 5.21–5.02 (m, 2H), 3.57 (s, 3H), 2.63 (s, 3H); ^{13}C NMR (151 MHz, CDCl$_3$): δ 197.4, 157.8, 140.1, 135.1, 130.8, 128.7, 128.6, 125.3, 125.2, 123.6, 123.0, 122.0, 122.0, 92.7, 70.4, 42.7, 27.1; LRMS (ESI): *m/z* 326.0 [M + H]$^+$; HRMS (ESI): calculated for C$_{18}$H$_{16}$NO$_3$S [M + H]$^+$: 326.0845, found: 326.0850.

Ethyl (5-methyl-5-oxido-8H-5λ^4-isochromeno[3,4-*c*][1,2]benzothiazin-2-yl)acetate (**3ja**): yellow solid; m.p.: 70–72 °C; ^{1}H NMR (600 MHz, CDCl$_3$): δ 7.96 (d, *J* = 1.8 Hz, 1H), 7.73 (d, *J* = 8.2 Hz, 1H), 7.59 (d, *J* = 7.9 Hz, 1H), 7.31 (td, *J* = 7.5, 1.6 Hz, 1H), 7.25 (d, *J* = 1.6 Hz, 1H), 7.19 (dd, *J* = 7.6, 1.5 Hz, 1H), 7.15 (td, *J* = 7.4, 1.1 Hz, 1H), 5.17–4.97 (m, 2H), 4.17 (q, *J* = 7.1 Hz, 2H), 3.68 (s, 2H), 3.48 (s, 3H), 1.27 (t, *J* = 7.1 Hz, 3H); ^{13}C NMR (126 MHz, CDCl$_3$): δ 170.7, 157.5, 139.4, 135.2, 131.2, 128.8, 128.2, 125.5, 125.0, 125.0, 124.8, 123.5, 122.1, 118.7, 91.8, 70.2, 61.3, 43.0, 41.7, 14.3; LRMS (ESI): *m/z* 370.0 [M + H]$^+$; HRMS (ESI): calculated for C$_{20}$H$_{20}$NO$_4$S [M + H]$^+$: 370.1108, found: 370.1116.

tert-Butyl (5-methyl-5-oxido-8H-5λ^4-isochromeno[3,4-*c*][1,2]benzothiazin-2-yl)carbamate (**3ka**): yellow-green solid; m.p.: 77–79 °C; ^{1}H NMR (600 MHz, CDCl$_3$): δ 7.90 (d, *J* = 2.0 Hz, 1H), 7.65 (d, *J* = 8.8 Hz, 1H), 7.59 (d, *J* = 7.4 Hz, 1H), 7.46 (d, *J* = 8.6 Hz, 1H), 7.30 – 7.24 (m, 1H), 7.13 (d, *J* = 7.5 Hz, 1H), 7.09 (t, *J* = 7.4 Hz, 1H), 6.98 (s, 1H), 5.08 – 4.95 (m, 2H), 3.39 (s, 3H), 1.49 (s, 9H); ^{13}C NMR (126 MHz, CDCl$_3$): δ 157.5, 152.3, 143.0, 136.3, 131.2, 128.6, 128.2, 124.9, 124.7, 124.6, 122.0, 115.1, 114.2, 111.9, 91.5, 81.4, 70.1, 43.4, 28.3; LRMS (ESI): *m/z* 399.0 [M + H]$^+$; HRMS (ESI): calculated for C$_{21}$H$_{23}$N$_2$O$_4$S [M + H]$^+$: 399.1373, found: 399.1383.

4,5-Dimethyl-8H-5λ^4-isochromeno[3,4-*c*][1,2]benzothiazine 5-oxide (**3la**): yellow solid; m.p.: 109–111 °C; ^{1}H NMR (600 MHz, CDCl$_3$): δ 7.90 (d, *J* = 6.5 Hz, 1H), 7.53 (d, *J* = 6.7 Hz, 1H), 7.44 (dd, *J* = 8.3, 7.3 Hz, 1H), 7.28 (td, *J* = 7.7, 1.6 Hz, 1H), 7.18 (dd, *J* = 7.4, 0.8 Hz, 1H), 7.13 (td, *J* = 7.4, 1.1 Hz, 1H), 7.10 (dt, *J* = 7.3, 1.0 Hz, 1H), 5.12–5.02 (m, 2H), 3.43 (s, 3H), 2.76 (s, 3H); ^{13}C NMR (126 MHz, CDCl$_3$): δ 156.2, 136.6, 135.4, 132.5, 131.7, 128.9, 128.1, 127.6, 125.0, 124.6, 123.0, 122.2, 120.0, 91.2, 70.2, 46.6, 21.1; LRMS (ESI): *m/z* 298.1 [M + H]$^+$; HRMS (ESI): calculated for C$_{17}$H$_{16}$NO$_2$S [M + H]$^+$: 298.0896, found: 298.0900.

4-Bromo-5-methyl-8H-5λ^4-isochromeno[3,4-*c*][1,2]benzothiazine 5-oxide (**3ma**): yellow solid; m.p.: 101–103 °C; ^{1}H NMR (500 MHz, CDCl$_3$): δ 8.07 (dd, *J* = 8.4, 1.1 Hz, 1H), 7.53–7.42 (m, 2H), 7.37 (t, *J* = 8.0 Hz, 1H), 7.31–7.25 (m, 1H), 7.19 (d, *J* = 7.6 Hz, 1H), 7.15 (t, *J* = 6.8 Hz, 1H), 5.19–5.02 (m, 2H), 3.89 (s, 3H); ^{13}C NMR (151 MHz, CDCl$_3$): δ 156.5, 138.5, 132.8, 131.1, 129.7, 128.9, 128.2, 125.2, 125.0, 124.6, 122.3, 120.9, 118.2, 91.2, 70.3, 49.2; LRMS (ESI): *m/z* 360.9 [M + H]$^+$; HRMS (ESI): calculated for C$_{16}$H$_{13}$BrNO$_2$S [M + H]$^+$: 361.9845, found: 361.9854.

3-Bromo-5-methyl-8H-5λ^4-isochromeno[3,4-*c*][1,2]benzothiazine 5-oxide (**3na**): yellow solid; m.p.: 120–122 °C; ^{1}H NMR (600 MHz, CDCl$_3$): δ 7.95 (d, *J* = 8.9 Hz, 1H), 7.88 (d, *J* = 2.1 Hz, 1H), 7.63 (dd, *J* = 8.9, 2.1 Hz, 1H), 7.52 (d, *J* = 8.3 Hz, 1H), 7.31 (td, *J* = 7.5, 1.7 Hz, 1H), 7.21–7.12 (m, 2H), 5.15–5.01 (m, 2H), 3.52 (s, 3H); ^{13}C NMR (151 MHz, CDCl$_3$): δ 157.3, 135.9, 133.7, 130.7, 128.7, 128.3, 126.2, 125.5, 125.1, 125.1, 122.1, 120.9, 116.2, 91.9, 70.3, 42.9; LRMS (ESI): *m/z* 361.2 [M + H]$^+$; HRMS (ESI): calculated for C$_{16}$H$_{13}$BrNO$_2$S [M + H]$^+$: 361.9845, found: 361.9843.

3-Methoxy-5-methyl-8H-5λ^4-isochromeno[3,4-*c*][1,2]benzothiazine 5-oxide (**3oa**) and 1-Methoxy-5-methyl-8H-5λ^4-isochromeno[3,4-*c*][1,2]benzothiazine 5-oxide (**3oa′**): pale yellow solid; ^{1}H NMR (500 MHz, CDCl$_3$) for the mixtures: δ 8.00 (d, *J* = 9.7 Hz, 1H), 7.54 (d, *J* = 7.9 Hz, 1H), 7.46 (d, *J* = 5.8 Hz, 1H), 7.41–7.33 (m, 5H), 7.30 (t, *J* = 7.4 Hz, 1H), 7.23–7.16 (m, 6H), 7.13 (d, *J* = 6.2 Hz, 7H), 7.08 (t, *J* = 7.2 Hz, 3H), 6.90 (d, *J* = 7.9 Hz, 1H), 6.85 (d, *J* = 7.8 Hz, 3H), 5.16–4.96 (m, 7H), 3.89 (s, 3H), 3.77 (s, 8H), 3.64 (s, 8H), 3.45 (s, 3H); ^{13}C NMR (126 MHz, CDCl$_3$) for the mixtures: δ 157.5, 154.8, 132.2, 132.0, 130.8, 128.2, 128.0, 127.6, 126.3, 125.8, 124.9, 124.5, 124.3, 124.1, 123.8, 123.2, 123.1, 123.1, 122.0, 121.4, 121.3, 120.0, 115.7, 115.1, 113.6, 113.5, 104.8, 89.4, 69.8, 69.6, 55.5, 54.8, 42.4, 42.0; LRMS (ESI): *m/z* 314.2 [M + H]$^+$; HRMS (ESI): calculated for C$_{17}$H$_{16}$NO$_3$S [M + H]$^+$: 314.0845, found: 314.0837.

8-Methyl-5H-8λ^4-isochromeno[3,4-*c*]naphtho[2,3-*e*][1,2]thiazine 8-oxide (**3pa**): yellow solid; m.p.: 98–100 °C; ^{1}H NMR (600 MHz, CDCl$_3$): δ 8.40 (s, 2H), 7.92 (d, *J* = 8.3 Hz, 1H), 7.82 (d, *J* = 8.4 Hz, 1H), 7.77 (d, *J* = 7.8 Hz, 1H), 7.61–7.55 (m, 1H), 7.46 (t, *J* = 7.5 Hz, 1H), 7.37 (t, *J* = 8.2 Hz, 1H), 7.23 (d, *J* = 7.1 Hz, 1H), 7.19 (t, *J* = 6.5 Hz, 1H), 5.11 (s, 2H), 3.41 (s, 3H); ^{13}C NMR (126 MHz, CDCl$_3$): δ 156.4, 135.8, 131.7, 130.2, 130.0, 129.3, 128.9, 128.8, 128.2, 127.8, 126.0, 125.1, 124.8, 124.6, 123.3, 121.9, 121.8, 91.8, 70.2, 42.5; LRMS (ESI): *m/z* 334.0 [M + H]$^+$; HRMS (ESI): calculated for C$_{20}$H$_{16}$NO$_2$S [M + H]$^+$: 334.0896, found: 334.0894.

5-Ethyl-8H-5λ^4-isochromeno[3,4-*c*][1,2]benzothiazine 5-oxide (**3qa**): yellow-green solid; m.p.: 77–79 °C; ^{1}H NMR (500 MHz, CDCl$_3$): δ 8.07 (d, *J* = 8.4 Hz, 1H), 7.72 (dd, *J* = 8.0, 1.4 Hz, 1H), 7.60 (d, *J* = 6.8 Hz, 1H), 7.59–7.55 (m, 1H), 7.34–7.27 (m, 2H), 7.19 (d, *J* = 5.9 Hz, 1H), 7.14 (td, *J* = 7.4, 1.2 Hz, 1H), 5.09 (q, *J* = 12.2 Hz, 2H), 3.60 (ddt, *J* = 32.6, 14.5, 7.3 Hz, 2H), 1.36 (t, *J* = 7.3 Hz, 3H); ^{13}C NMR (126 MHz, CDCl$_3$): δ 157.7, 136.0, 133.0, 131.4, 128.8, 128.2, 125.0, 124.7, 124.6, 124.2, 123.8, 122.0, 117.2, 91.3, 70.2,

49.3, 7.9; LRMS (ESI): m/z 298.1 [M + H]$^+$; HRMS (ESI): calculated for $C_{17}H_{16}NO_2S$ [M + H]$^+$: 298.0896, found: 298.0904.

5-Cyclopropyl-8H-5λ^4-isochromeno[3,4-c][1,2]benzothiazine 5-oxide (**3ra**): yellow oil; ^{1}H NMR (600 MHz, CDCl$_3$): δ 8.09 (d, J = 8.3 Hz, 1H), 7.87 (d, J = 8.0 Hz, 1H), 7.62 (d, J = 7.8 Hz, 1H), 7.56 (t, J = 7.7 Hz, 1H), 7.33–7.28 (m, 2H), 7.19 (d, J = 7.3 Hz, 1H), 7.14 (t, J = 7.4 Hz, 1H), 5.17–4.99 (m, 2H), 2.88 (tt, J = 8.0, 4.7 Hz, 1H), 1.78 – 1.70 (m, 1H), 1.57–1.49 (m, 1H), 1.42 (dt, J = 14.9, 7.8 Hz, 1H), 1.31–1.26 (m, 1H); ^{13}C NMR (151 MHz, CDCl$_3$): δ 157.4, 135.2, 132.5, 131.3, 128.8, 128.1, 125.0, 124.7, 124.4, 124.1, 123.3, 122.3, 120.5, 91.8, 70.2, 30.7, 7.0, 4.6; LRMS (ESI): m/z 310.1 [M + H]$^+$; HRMS (ESI): calculated for $C_{18}H_{16}NO_2S$ [M + H]$^+$: 310.0896, found: 310.0904.

5-(Chloromethyl)-8H-5λ^4-isochromeno[3,4-c][1,2]benzothiazine 5-oxide (**3sa**): yellow-green solid; m.p.: 194–196 °C; ^{1}H NMR (600 MHz, CDCl$_3$): δ 8.06 (d, J = 8.4 Hz, 1H), 7.87 (dd, J = 8.1, 1.4 Hz, 1H), 7.66–7.62 (m, 1H), 7.61 (d, J = 7.4 Hz, 1H), 7.38–7.34 (m, 1H), 7.32 (td, J = 7.5, 1.7 Hz, 1H), 7.21 (dd, J = 7.6, 1.6 Hz, 1H), 7.17 (td, J = 7.3, 1.1 Hz, 1H), 5.09 (q, J = 12.3 Hz, 2H), 4.89–4.75 (m, 2H); ^{13}C NMR (126 MHz, CDCl$_3$): δ 157.3, 137.3, 134.3, 130.8, 128.9, 128.3, 125.8, 125.1, 125.1, 124.7, 124.4, 122.1, 114.6, 91.7, 70.4, 58.6; LRMS (ESI): m/z 317.9 [M + H]$^+$; HRMS (ESI): calculated for $C_{16}H_{13}ClNO_2S$ [M + H]$^+$: 318.0350, found: 318.0345.

2-(5-Oxido-8H-5λ^4-isochromeno[3,4-c][1,2]benzothiazin-5-yl)ethanol (**3ta**): yellow-green solid; m.p.: 143–145 °C; ^{1}H NMR (600 MHz, CDCl$_3$): δ 8.08 (d, J = 8.3 Hz, 1H), 7.80 (dd, J = 8.1, 1.3 Hz, 1H), 7.64–7.56 (m, 2H), 7.36–7.29 (m, 2H), 7.21–7.14 (m, 2H), 5.19–5.03 (m, 2H), 4.18–4.07 (m, 2H), 3.95 – 3.83 (m, 2H), 3.26 (s, 1H); ^{13}C NMR (151 MHz, CDCl$_3$): δ 156.5, 135.1, 132.9, 130.9, 129.7, 128.6, 128.1, 126.4, 125.0, 124.4, 123.4, 122.2, 118.8, 92.2, 70.2, 56.3, 56.2; LRMS (ESI): m/z 314.1 [M + H]$^+$; HRMS (ESI): calculated for $C_{17}H_{16}NO_3S$ [M + H]$^+$: 314.0845, found: 314.0842.

5-Phenyl-8H-5λ^4-isochromeno[3,4-c][1,2]benzothiazine 5-oxide (**3ua**): yellow solid; m.p.: 102–103 °C; ^{1}H NMR (600 MHz, CDCl$_3$): δ 8.13 (d, J = 8.4 Hz, 1H), 8.07 (d, J = 7.8 Hz, 2H), 7.71 (t, J = 7.4 Hz, 1H), 7.67 (d, J = 7.8 Hz, 1H), 7.62 (t, J = 7.7 Hz, 2H), 7.51 (t, J = 7.7 Hz, 1H), 7.34 (t, J = 7.6 Hz, 1H), 7.22 (t, J = 8.0 Hz, 2H), 7.19–7.13 (m, 2H), 5.24–5.10 (m, 2H); ^{13}C NMR (151 MHz, CDCl$_3$): δ 157.4, 136.8, 134.6, 134.3, 132.2, 131.3, 129.8, 129.3, 128.9, 128.2, 125.0, 124.9, 124.7, 124.2, 124.1, 122.5, 121.2, 92.1, 70.4; LRMS (ESI): m/z 346.0 [M + H]$^+$; HRMS (ESI): calculated for $C_{21}H_{16}NO_2S$ [M + H]$^+$: 346.0896, found: 346.0899.

5-Benzyl-8H-5λ^4-isochromeno[3,4-c][1,2]benzothiazine 5-oxide (**3va**): yellow-green solid; m.p.: 66–68 °C; ^{1}H NMR (600 MHz, CDCl$_3$): δ 7.99 (dd, J = 8.1, 1.1 Hz, 1H), 7.75 (dd, J = 7.8, 1.4 Hz, 1H), 7.65 (d, J = 8.9 Hz, 1H), 7.59 – 7.53 (m, 1H), 7.38 (td, J = 7.5, 1.1 Hz, 1H), 7.32 (t, J = 6.9 Hz, 1H), 7.30–7.26 (m, 2H), 7.25–7.22 (m, 1H), 7.18 (td, J = 7.4, 1.1 Hz, 1H), 7.16–7.12 (m, 3H), 5.33 (d, J = 15.5 Hz, 1H), 5.08 (d, J = 12.1 Hz, 1H), 4.98 (d, J = 15.5 Hz, 1H), 4.62 (d, J = 12.1 Hz, 1H); ^{13}C NMR (126 MHz, CDCl$_3$): δ 147.2, 137.0, 133.9, 131.4, 130.6, 129.0, 128.8, 128.5, 128.3, 127.8, 127.4, 126.5, 125.8, 125.4, 125.3, 124.9, 123.5, 97.0, 70.4, 54.2; LRMS (ESI): m/z 360.0 [M + H]$^+$; HRMS (ESI): calculated for $C_{22}H_{18}NO_2S$ [M + H]$^+$: 360.1053, found: 360.1053.

5,10-Dimethyl-8H-5λ^4-isochromeno[3,4-c][1,2]benzothiazine 5-oxide (**3ab**): yellow-green solid; m.p.: 152–154 °C; ^{1}H NMR (600 MHz, CDCl$_3$): δ 8.05 (d, J = 8.4 Hz, 1H), 7.77 (d, J = 7.7 Hz, 1H), 7.59–7.54 (m, 1H), 7.49 (d, J = 7.9 Hz, 1H), 7.32 (t, J = 7.6 Hz, 1H), 7.12 (dd, J = 8.0, 1.8 Hz, 1H), 7.01 (s, 1H), 5.12–5.00 (m, 2H), 3.49 (s, 3H), 2.36 (s, 3H); ^{13}C NMR (151 MHz, CDCl$_3$): δ 156.8, 135.1, 134.6, 132.8, 129.0, 128.8, 128.4, 125.7, 124.5, 124.3, 123.2, 122.2, 119.9, 91.9, 70.3, 43.0, 21.1.; LRMS (ESI): m/z 298.1 [M + H]$^+$; HRMS (ESI): calculated for $C_{17}H_{16}NO_2S$ [M + H]$^+$: 298.0896, found: 298.0902.

10-Methoxy-5-methyl-8H-5λ^4-isochromeno[3,4-c][1,2]benzothiazine 5-oxide (**3ac**): yellow-green solid; m.p.: 108–110 °C; ^{1}H NMR (600 MHz, CDCl$_3$): δ 8.02 (d, J = 8.4 Hz, 1H), 7.76 (d, J = 8.1 Hz, 1H), 7.56 (t, J = 7.7 Hz, 1H), 7.51 (d, J = 8.5 Hz, 1H), 7.31 (t, J = 7.6 Hz, 1H), 6.87 (dd, J = 8.6, 2.7 Hz, 1H), 6.76 (d, J = 2.7 Hz, 1H), 5.05 (q, J = 12.3 Hz, 2H), 3.82 (s, 3H), 3.50 (s, 3H); ^{13}C NMR (151 MHz, CDCl$_3$):

δ 157.2, 156.2, 135.0, 132.8, 130.5, 124.4, 124.2, 123.9, 123.5, 123.1, 119.8, 113.5, 110.9, 91.7, 70.2, 55.6, 42.9; LRMS (ESI): *m/z* 314.1 [M + H]$^+$; HRMS (ESI): calculated for $C_{17}H_{16}NO_3S$ [M + H]$^+$: 314.0845, found: 314.0847.

11-Methoxy-5-methyl-8H-5λ^4-isochromeno[3,4-*c*][1,2]benzothiazine 5-oxide (**3ad**): yellow solid; m.p.: 209–211 °C; ^{1}H NMR (600 MHz, CDCl$_3$): δ 8.06 (d, *J* = 8.3 Hz, 1H), 7.74 (d, *J* = 6.7 Hz, 1H), 7.60–7.51 (m, 1H), 7.32–7.27 (m, 1H), 7.12 (d, *J* = 2.5 Hz, 1H), 7.08 (d, *J* = 8.2 Hz, 1H), 6.66 (dd, *J* = 8.2, 2.5 Hz, 1H), 5.07–4.95 (m, 2H), 3.78 (s, 3H), 3.46 (s, 3H); ^{13}C NMR (151 MHz, CDCl$_3$): δ 159.7, 157.5, 135.0, 133.0, 132.6, 125.9, 124.5, 124.4, 123.2, 121.4, 120.0, 109.4, 108.7, 91.9, 69.9, 55.5, 42.9; LRMS (ESI): *m/z* 314.0 [M + H]$^+$; HRMS (ESI): calculated for $C_{17}H_{16}NO_3S$ [M + H]$^+$: 314.0845, found: 314.0842.

10-Chloro-5-methyl-8H-5λ^4-isochromeno[3,4-*c*][1,2]benzothiazine 5-oxide (**3ae**): yellow solid; m.p.: 180–182 °C; ^{1}H NMR (500 MHz, CDCl$_3$): δ 7.98 (d, *J* = 7.3 Hz, 1H), 7.78 (dd, *J* = 8.1, 1.3 Hz, 1H), 7.60–7.56 (m, 1H), 7.51 (d, *J* = 8.3 Hz, 1H), 7.37–7.31 (m, 1H), 7.28–7.24 (m, 1H), 7.17 (d, *J* = 2.2 Hz, 1H), 5.11–4.95 (m, 2H), 3.51 (s, 3H); ^{13}C NMR (126 MHz, CDCl$_3$): δ 157.3, 134.7, 133.0, 130.4, 129.9, 129.8, 128.2, 125.1, 124.6, 124.3, 123.3, 123.3, 120.1, 91.3, 69.6, 43.0; LRMS (ESI): *m/z* 318.1 [M + H]$^+$; HRMS (ESI): calculated for $C_{16}H_{13}ClNO_2S$ [M + H]$^+$: 318.0350, found: 318.0359.

11-Chloro-5-methyl-8H-5λ^4-isochromeno[3,4-*c*][1,2]benzothiazine 5-oxide (**3af**): yellow-green solid; m.p.: 66–68 °C; ^{1}H NMR (600 MHz, CDCl$_3$): δ 8.02 (d, *J* = 8.2 Hz, 1H), 7.78 (dd, *J* = 8.1, 1.3 Hz, 1H), 7.65–7.60 (m, 1H), 7.56 (s, 1H), 7.38–7.33 (m, 1H), 7.11 (d, *J* = 1.3 Hz, 2H), 5.13–4.95 (m, 2H), 3.50 (s, 3H); ^{13}C NMR (151 MHz, CDCl$_3$): δ 157.7, 134.6, 134.2, 133.3, 133.2, 126.9, 126.2, 124.8, 124.5, 124.2, 123.3, 121.9, 120.1, 91.2, 69.7, 43.0; LRMS (ESI): *m/z* 318.1 [M + H]$^+$; HRMS (ESI): calculated for $C_{16}H_{13}ClNO_2S$ [M + H]$^+$: 318.0350, found: 318.0349.

10-Fluoro-5-methyl-8H-5λ^4-isochromeno[3,4-*c*][1,2]benzothiazine 5-oxide (**3ag**): yellow-green solid; m.p.: 77–79 °C; ^{1}H NMR (600 MHz, CDCl$_3$): δ 8.00 (d, *J* = 8.4 Hz, 1H), 7.78 (dd, *J* = 8.1, 1.3 Hz, 1H), 7.60–7.55 (m, 1H), 7.53 (dd, *J* = 8.6, 5.2 Hz, 1H), 7.40–7.30 (m, 1H), 7.00 (td, *J* = 8.6, 2.8 Hz, 1H), 6.91 (dd, *J* = 8.4, 2.7 Hz, 1H), 5.12–4.97 (m, 2H), 3.52 (s, 3H); ^{13}C NMR (151 MHz, CDCl$_3$): δ 161.1, 159.5, 156.8, 134.7, 132.9, 130.7 (d, *J* = 7.0 Hz), 127.3 (d, *J* = 3.2 Hz), 124.4 (d, *J* = 42.3 Hz), 123.6 (d, *J* = 7.7 Hz), 123.2, 120.1, 114.9 (d, *J* = 21.4 Hz), 112.3 (d, *J* = 22.3 Hz), 91.4, 69.6 (d, *J* = 2.1 Hz), 42.9; LRMS (ESI): *m/z* 302.1 [M + H]$^+$; HRMS (ESI): calculated for $C_{16}H_{13}FNO_2S$ [M + H]$^+$: 302.0646, found: 302.0649.

11-Fluoro-5-methyl-8H-5λ^4-isochromeno[3,4-*c*][1,2]benzothiazine 5-oxide (**3ah**): pale yellow solid; m.p.: 91–93 °C; ^{1}H NMR (600 MHz, CDCl$_3$): δ 8.02 (d, *J* = 8.9 Hz, 1H), 7.78 (dd, *J* = 8.1, 1.3 Hz, 1H), 7.67–7.57 (m, 1H), 7.38–7.32 (m, 1H), 7.28 (dd, *J* = 10.6, 2.5 Hz, 1H), 7.14 (dd, *J* = 8.3, 5.7 Hz, 1H), 6.82 (td, *J* = 8.4, 2.5 Hz, 1H), 5.12–4.97 (m, 2H), 3.51 (s, 3H); ^{13}C NMR (151 MHz, CDCl$_3$): δ 163.7, 162.1, 157.6, 134.5, 133.3 (d, *J* = 9.2 Hz), 133.1, 126.3 (d, *J* = 9.3 Hz), 124.6, 124.1 (d, *J* = 3.6 Hz), 123.2, 120.0, 111.0 (d, *J* = 22.3 Hz), 109.0 (d, *J* = 24.1 Hz), 91.3 (d, *J* = 2.2 Hz), 69.6, 42.9; LRMS (ESI): *m/z* 302.0 [M + H]$^+$; HRMS (ESI): calculated for $C_{16}H_{13}FNO_2S$ [M + H]$^+$: 302.0646, found: 302.0645.

5-Methyl-10-(trifluoromethyl)-8H-5λ^4-isochromeno[3,4-*c*][1,2]benzothiazine 5-oxide (**3ai**): yellow solid; m.p.: 191–193 °C; ^{1}H NMR (600 MHz, CDCl$_3$): δ 8.01 (d, *J* = 8.3 Hz, 1H), 7.80 (dd, *J* = 8.0, 1.3 Hz, 1H), 7.68 (d, *J* = 8.2 Hz, 1H), 7.63–7.58 (m, 1H), 7.56–7.52 (m, 1H), 7.44 (s, 1H), 7.42–7.35 (m, 1H), 5.21–5.02 (m, 2H), 3.53 (s, 3H); ^{13}C NMR (151 MHz, CDCl$_3$): δ 158.3, 135.0, 134.5, 133.2, 128.7, 126.4 (q, *J* = 32.6 Hz), 125.3 (d, *J* = 3.6 Hz), 124.9, 124.3, 123.3, 122.0 (d, *J* = 3.9 Hz), 121.9, 120.2, 91.4, 69.7, 43.0; LRMS (ESI): *m/z* 352.0 [M + H]$^+$; HRMS (ESI): calculated for $C_{17}H_{13}F_3NO_2S$ [M + H]$^+$: 352.0614, found: 352.0617.

5-Methyl-11-(trifluoromethyl)-8H-5λ^4-isochromeno[3,4-*c*][1,2]benzothiazine 5-oxide (**3aj**): yellow solid; m.p.: 150–152 °C; ^{1}H NMR (600 MHz, CDCl$_3$): δ 8.00 (d, *J* = 8.3 Hz, 1H), 7.83 (s, 1H), 7.81 (dd, *J* = 8.1, 1.3 Hz, 1H), 7.67–7.61 (m, 1H), 7.42–7.36 (m, 2H), 7.30 (d, *J* = 7.8 Hz, 1H), 5.20–5.04 (m, 2H), 3.52 (s, 3H); ^{13}C NMR (151 MHz, CDCl$_3$): δ 157.8, 134.5, 133.5, 132.2, 131.9, 131.0–130.3 (m), 125.5, 125.0, 124.0,

123.4, 121.5–121.4 (m), 120.2, 118.6 (q, J = 3.9 Hz), 91.3, 69.7, 43.0; LRMS (ESI): m/z 352.0 [M + H]$^+$; HRMS (ESI): calculated for $C_{17}H_{13}F_3NO_2S$ [M + H]$^+$: 352.0614, found: 352.0618.

3.2.5. Mechanistic Investigations

KIE Experiment. To a 15 mL vial was added d^1-**1a** (23.43 mg, 0.15 mmol), **2a** (54.01 mg, 0.165 mmol), [Cp*RhCl$_2$]$_2$ (4.64 mg, 5 mol%), AgOPiv (6.27 mg, 20 mol%) under air. Trifluoroethanol (TFE, 2.5 mL) was added subsequently. The resulting mixture was stirred at ambient temperature for 6 h. Then, the mixture was filtered through a celite pad and washed with DCM (10 mL × 3). The combined organic layer was concentrated under vacuo, and the residue was purified by silica gel chromatography eluting with DCM/MeOH from 50:1 to 30:1 to give the mixture of **3aa** and d^1-**3aa**. The ratio of two products was determined by ^{1}H NMR to give an intramolecular kinetic isotopic effect (KIE) kH/kD = 0.47. (see Supporting Information).

H/D Exchange Experiment. A mixture of **1a** (23.43 mg, 0.15 mmol), [Cp*RhCl$_2$]$_2$ (4.64 mg, 5 mol%), AgOPiv (6.27 mg, 20 mol%) and d^3-TFE was added into a vial under air. The resulting mixture was stirred at room temperature for 18 h. Then, the mixture was filtered through a celite pad and washed with DCM (10 mL × 3). The combined organic layer was concentrated under vacuo, and the residue was purified by silica gel chromatography eluting with DCM/MeOH 10:1 to give the product d^2-**1a**. H/D exchange occurred at the o-position of S-phenylsulfoximine (91% D). (see Supporting Information).

Competition Experiment. To a mixture of **1c** (27.79 mg, 0.15 mmol), **1i** (29.59 mg, 0.15 mmol), **2a** (54.01 mg, 0.165 mmol), [Cp*RhCl$_2$]$_2$ (4.64 mg, 5 mol%), and AgOPiv (6.27 mg, 20 mol%) was added TFE (2.5 mL) under air. The resulting mixture was stirred at room temperature for 18 h. After the reaction was completed, the mixture was filtered through a celite pad and washed with DCM (10 mL × 3). The combined organic layer was concentrated under vacuo and the residue was purified by silica gel chromatography eluting with DCM/MeOH from 100:1 to 30:1 to give the isolated products **3ca** and **3ia**. The ratio was calculated according to the moles of products.

4. Conclusions

In summary, we have developed a sulfoximie-assisted Rh(III)-catalyzed C–H activation and intramolecular annulation. In this strategy, fused isochromeno-1,2-benzothiazines was unprecedentedly synthesized, and the desired products could be yields in good to excellent yields. Most importantly, it is a redox-neutral process that can be conducted under room temperature and the strategy features broad generality and versatility.

Author Contributions: Experiments and investigation, B.W., X.H. and J.L.; formal analysis and data curation, B.W.; writing—original draft preparation, B.W. and X.H.; writing—review and editing, C.L. and H.L. All authors have read and agreed to the published version of the manuscript.

Funding: We are grateful to the National Natural Science Foundation of China (No. 21672232, 21977106, 81620108027), National S&T Major Projects (2018ZX09711002) and the Strategic Priority Research Program of the Chinese Academy of Sciences (XDA12040217) for financial support.

Conflicts of Interest: The authors declare no conflict of interest.

References

1. Cho, G.Y.; Bolm, C. Silver-Catalyzed Imination of Sulfoxides and Sulfides. *Org. Lett.* **2005**, *7*, 4983–4985. [CrossRef]
2. Gais, H.-J. Development of New Methods for Asymmetric Synthesis Based on Sulfoximines. *Heteroat. Chem.* **2007**, *18*, 472–481. [CrossRef]
3. Hendriks, C.M.M.; Lamers, P.; Engel, J.; Bolm, C. Sulfoxide-to-Sulfilimine Conversions: Use of Modified Burgess-Type Reagents. *Adv. Synth. Catal.* **2013**, *355*, 3363–3368. [CrossRef]
4. Bizet, V.; Kowalczyk, R.; Bolm, C. Fluorinated sulfoximines: Syntheses, properties and applications. *Chem. Soc. Rev.* **2014**, *43*, 2426–2438. [CrossRef] [PubMed]

5. Miao, J.; Richards, N.G.J.; Ge, H. Rhodium-catalyzed direct synthesis of unprotected NH-sulfoximines from sulfoxides. *Chem. Commun.* **2014**, *50*, 9687–9689. [CrossRef] [PubMed]
6. Bizet, V.; Hendriks, C.M.M.; Bolm, C. Sulfur imidations: Access to sulfimides and sulfoximines. *Chem. Soc. Rev.* **2015**, *44*, 3378–3390. [CrossRef] [PubMed]
7. Tota, A.; Zenzola, M.; Chawner, S.J.; John-Campbell, S.S.; Carlucci, C.; Romanazzi, G.; DeGennaro, L.; Bull, J.A.; Luisi, R. Synthesis of NH-sulfoximines from sulfides by chemoselective one-pot N- and O-transfers. *Chem. Commun.* **2017**, *53*, 348–351. [CrossRef]
8. Davies, T.; Hall, A.; Willis, M.C. One-Pot, Three-Component Sulfonimidamide Synthesis Exploiting the Sulfinylamine Reagent N -Sulfinyltritylamine, TrNSO. *Angew. Chem. Int. Ed.* **2017**, *56*, 14937–14941. [CrossRef]
9. Yu, H.; Li, Z.; Bolm, C. Iron(II)-Catalyzed Direct Synthesis of NH Sulfoximines from Sulfoxides. *Angew. Chem. Int. Ed.* **2017**, *57*, 324–327. [CrossRef]
10. Lücking, U. Sulfoximines: A Neglected Opportunity in Medicinal Chemistry. *Angew. Chem. Int. Ed.* **2013**, *52*, 9399–9408. [CrossRef]
11. MarcusFrings; CarstenBolm; AndreasBlum; ChristianGnamm, Sulfoximines from a Medicinal Chemist's Perspective: Physicochemical and in vitro Parameters Relevant for Drug Discovery. *Eur. J. Med. Chem.* **2017**, *126*, 225–245. [CrossRef] [PubMed]
12. Sirvent, J.A.; Lücking, U. Novel Pieces for the Emerging Picture of Sulfoximines in Drug Discovery: Synthesis and Evaluation of Sulfoximine Analogues of Marketed Drugs and Advanced Clinical Candidates. *ChemMedChem* **2017**, *12*, 487–501. [CrossRef]
13. Lücking, U. Neglected sulfur(vi) pharmacophores in drug discovery: Exploration of novel chemical space by the interplay of drug design and method development. *Org. Chem. Front.* **2019**, *6*, 1319–1324. [CrossRef]
14. Dillard, R.D.; Yen, T.T.; Stark, P.; Pavey, D.E. Synthesis and Blood Pressure Lowering Activity of 3-(Substituted-amino)-1,2,4-benzothiadiazine 1-Oxide Derivatives. *J. Med. Chem.* **1980**, *23*, 717–722. [CrossRef]
15. Buckheit, R.W.; Fliakas-Boltz, V.; Decker, W.; Roberson, J.L.; Pyle, C.A.; White, E.; Bowdon, B.J.; McMahon, J.B.; Boyd, M.R.; Bader, J.P.; et al. Biological and biochemical anti-HIV activity of the benzothiadiazine class of nonnucleoside reverse transcriptase inhibitors. *Antivir. Res.* **1994**, *25*, 43–56. [CrossRef]
16. Okamura, H.; Bolm, C. Rhodium-Catalyzed Imination of Sulfoxides and Sulfides: Efficient Preparation of N-Unsubstituted Sulfoximines and Sulfilimines. *Org. Lett.* **2004**, *6*, 1305–1307. [CrossRef] [PubMed]
17. Xie, Y.; Zhou, B.; Zhou, S.; Wei, W.; Liu, J.; Zhan, Y.; Cheng, D.; Chen, M.; Li, Y.; Wang, B.; et al. Sulfimine-Promoted Fast O Transfer: One-step Synthesis of Sulfoximine from Sulfide. *ChemistrySelect* **2017**, *2*, 1620–1624. [CrossRef]
18. Alberico, D.; Scott, M.E.; Lautens, M. Aryl–Aryl Bond Formation by Transition-Metal-Catalyzed Direct Arylation. *Chem. Rev.* **2007**, *107*, 174–238. [CrossRef]
19. Chen, X.; Engle, K.M.; Wang, N.-H.; Yu, J.-Q. ChemInform Abstract: Palladium(II)-Catalyzed C-H Activation/C-C Cross-Coupling Reactions: Versatility and Practicality. *Chemin* **2009**, *40*, 5094–5115. [CrossRef]
20. Cho, S.H.; Kim, J.Y.; Kwak, J.; Chang, S. ChemInform Abstract: Recent Advances in the Transition Metal-Catalyzed Twofold Oxidative C-H Bond Activation Strategy for C-C and C-N Bond Formation. *Chemin* **2011**, *43*, 5068–5083. [CrossRef]
21. Colby, D.A.; Tsai, A.S.; Bergman, R.G.; Ellman, J.A. Rhodium Catalyzed Chelation-Assisted C–H Bond Functionalization Reactions. *Accounts Chem. Res.* **2011**, *45*, 814–825. [CrossRef]
22. Song, G.; Wang, F.; Li, X. ChemInform Abstract: C-C, C-O and C-N Bond Formation via Rhodium(III)-Catalyzed Oxidative C-H Activation. *Chemin* **2012**, *43*, 3651–3678. [CrossRef]
23. Rouquet, G.; Chatani, N. Catalytic Functionalization of C(sp 2)?H and C(sp 3)?H Bonds by Using Bidentate Directing Groups. *Angew. Chem. Int. Ed.* **2013**, *52*, 11726–11743. [CrossRef]
24. Wencel-Delord, J.; Glorius, F. ChemInform Abstract: C-H Bond Activation Enables the Rapid Construction and Late-Stage Diversification of Functional Molecules. *Chemin* **2013**, *44*, 369–375. [CrossRef]
25. Song, G.; Li, X. Substrate Activation Strategies in Rhodium(III)-Catalyzed Selective Functionalization of Arenes. *Acc. Chem. Res.* **2015**, *48*, 1007–1020. [CrossRef]
26. Hummel, J.R.; Boerth, J.A.; Ellman, J.A. Transition-Metal-Catalyzed C–H Bond Addition to Carbonyls, Imines, and Related Polarized π Bonds. *Chem. Rev.* **2016**, *117*, 9163–9227. [CrossRef]
27. Xia, Y.; Qiu, D.; Wang, J. Transition-Metal-Catalyzed Cross-Couplings through Carbene Migratory Insertion. *Chem. Rev.* **2017**, *117*, 13810–13889. [CrossRef]

28. Fernández-Rodríguez, M.A.; Shen, Q.; Hartwig, J.F. A General and Long-Lived Catalyst for the Palladium-Catalyzed Coupling of Aryl Halides with Thiols. *J. Am. Chem. Soc.* **2006**, *128*, 2180–2181.
29. Platon, M.; Wijaya, N.; Rampazzi, V.; Cui, L.; Rousselin, Y.; Saeys, M.; Hierso, J.-C. Thioetherification of Chloroheteroarenes: A Binuclear Catalyst Promotes Wide Scope and High Functional-Group Tolerance. *Chem. A Eur. J.* **2014**, *20*, 12584–12594. [CrossRef]
30. Guilbaud, J.; Labonde, M.; Selmi, A.; Kammoun, M.; Cattey, H.; Pirio, N.; Roger, J.; Hierso, J.-C. Palladium-catalyzed heteroaryl thioethers synthesis overcoming palladium dithiolate resting states inertness: Practical road to sulfones and NH-sulfoximines. *Catal. Commun.* **2018**, *111*, 52–58. [CrossRef]
31. Yu, D.-G.; De Azambuja, F.; Glorius, F. α-MsO/TsO/Cl Ketones as Oxidized Alkyne Equivalents: Redox-Neutral Rhodium(III)-Catalyzed C-H Activation for the Synthesis of N-Heterocycles. *Angew. Chem. Int. Ed.* **2014**, *53*, 2754–2758. [CrossRef] [PubMed]
32. Wen, J.; Tiwari, D.P.; Bolm, C. 1,2-Benzothiazines from Sulfoximines and Allyl Methyl Carbonate by Rhodium-Catalyzed Cross-Coupling and Oxidative Cyclization. *Org. Lett.* **2017**, *19*, 1706–1709. [CrossRef] [PubMed]
33. Xie, H.; Lan, J.; Gui, J.; Chen, F.; Jiang, H.; Zeng, W.; Jiang, H. Ru (II)-Catalyzed Coupling-Cyclization of Sulfoximines with alpha -Carbonyl Sulfoxonium Ylides as an Approach to 1,2-Benzothiazines. *Adv. Synth. Catal.* **2018**, *360*, 3534–3543. [CrossRef]
34. Dong, W.; Wang, L.; Parthasarathy, K.; Pan, F.; Bolm, C. Rhodium-Catalyzed Oxidative Annulation of Sulfoximines and Alkynes as an Approach to 1,2-Benzothiazines. *Angew. Chem. Int. Ed.* **2013**, *52*, 11573–11576. [CrossRef]
35. Cheng, Y.; Bolm, C. Regioselective Syntheses of 1,2-Benzothiazines by Rhodium-Catalyzed Annulation Reactions. *Angew. Chem. Int. Ed.* **2015**, *54*, 12349–12352. [CrossRef]
36. Jeon, W.H.; Son, J.-Y.; Kim, J.E.; Lee, P.H. ChemInform Abstract: Synthesis of 1,2-Benzothiazines by a Rhodium-Catalyzed Domino C-H Activation/Cyclization/Elimination Process from S-Aryl Sulfoximines and Pyridotriazoles. *Chemin* **2016**, *47*, 3498–3501. [CrossRef]
37. Liu, Y.-Z.; Hu, Y.; Lv, G.-H.; Nie, R.-F.; Peng, Y.; Zhang, C.; Lv, S.-Y.; Hai, L.; Wang, H.-J.; Wu, Y. Synthesis of 1,2-Benzothiazines via C–H Activation/Cyclization in a Recyclable, Mild System. *ACS Sustain. Chem. Eng.* **2019**, *7*, 13425–13429. [CrossRef]
38. Ko, G.H.; Son, J.-Y.; Kim, H.; Maeng, C.; Baek, Y.; Seo, B.; Um, K.; Lee, P.H. Synthesis of Indolo-1,2-Benzothiazines from Sulfoximines and 3-Diazoindolin-2-imines. *Adv. Synth. Catal.* **2017**, *359*, 3362–3370. [CrossRef]
39. Wu, X.; Wang, B.; Zhou, S.; Zhou, Y.; Liu, H. Ruthenium-Catalyzed Redox-Neutral [4 + 1] Annulation of Benzamides and Propargyl Alcohols via C–H Bond Activation. *ACS Catal.* **2017**, *7*, 2494–2499. [CrossRef]
40. Xie, Y.; Wu, X.; Cai, J.; Wang, J.; Li, J.; Liu, H. Ruthenium(II)-Catalyzed Redox-Neutral [3+2] Annulation of Indoles with Internal Alkynes via C–H Bond Activation: Accessing a Pyrroloindolone Scaffold. *J. Org. Chem.* **2017**, *82*, 5263–5273. [CrossRef]
41. Han, X.; Gao, F.; Li, C.; Fang, D.; Xie, X.; Zhou, Y.; Liu, H. Synthesis of Highly Fused Pyrano[2,3-b]pyridines via Rh(III)-Catalyzed C–H Activation and Intramolecular Cascade Annulation under Room Temperature. *J. Org. Chem.* **2020**, *85*, 6281–6294. [CrossRef] [PubMed]
42. Back, T.G. Organosulfur Chemistry in Asymmetric Synthesis. *Angew. Chem.* **2009**, *121*, 2112–2114. [CrossRef]
43. Otocka, S.; Kwiatkowska, M.; Madalińska, L.; Kiełbasiński, P. Chiral Organosulfur Ligands/Catalysts with a Stereogenic Sulfur Atom: Applications in Asymmetric Synthesis. *Chem. Rev.* **2017**, *117*, 4147–4181. [CrossRef] [PubMed]
44. Shen, B.; Wan, B.; Li, X. Enantiodivergent Desymmetrization in the Rhodium(III)-Catalyzed Annulation of Sulfoximines with Diazo Compounds. *Angew. Chem. Int. Ed.* **2018**, *57*, 15534–15538. [CrossRef]
45. Sun, Y.; Cramer, N. Enantioselective Synthesis of Chiral-at-Sulfur 1,2-Benzothiazines by CpXRhIII-Catalyzed C-H Functionalization of Sulfoximines. *Angew. Chem. Int. Ed.* **2018**, *57*, 15539–15543. [CrossRef]
46. Brauns, M.; Cramer, N. Efficient Kinetic Resolution of Sulfur-Stereogenic Sulfoximines by Exploiting CpXRhIII -Catalyzed C-H Functionalization. *Angew. Chem. Int. Ed.* **2019**, *58*, 8902–8906. [CrossRef]
47. Simmons, E.M.; Hartwig, J.F. On the Interpretation of Deuterium Kinetic Isotope Effects in C?H Bond Functionalizations by Transition-Metal Complexes. *Angew. Chem. Int. Ed.* **2012**, *51*, 3066–3072. [CrossRef]

48. Wang, H.; Li, L.; Yu, S.; Li, Y.; Li, X. ChemInform Abstract: Rh(III)-Catalyzed C-C/C-N Coupling of Imidates with α-Diazo Imidamide: Synthesis of Isoquinoline-Fused Indoles. *Chemin* **2016**, *47*, 2914–2917. [CrossRef]
49. Zhu, R.-Y.; He, J.; Wang, X.-C.; Yu, J.-Q. Ligand-Promoted Alkylation of C(sp3)–H and C(sp2)–H Bonds. *J. Am. Chem. Soc.* **2014**, *136*, 13194–13197. [CrossRef]

Sample Availability: Samples of the compounds are not available from the authors.

© 2020 by the authors. Licensee MDPI, Basel, Switzerland. This article is an open access article distributed under the terms and conditions of the Creative Commons Attribution (CC BY) license (http://creativecommons.org/licenses/by/4.0/).

MDPI
St. Alban-Anlage 66
4052 Basel
Switzerland
Tel. +41 61 683 77 34
Fax +41 61 302 89 18
www.mdpi.com

Molecules Editorial Office
E-mail: molecules@mdpi.com
www.mdpi.com/journal/molecules

www.ingramcontent.com/pod-product-compliance
Lightning Source LLC
LaVergne TN
LVHW070605100526
838202LV00012B/565